Unstable Singularities and Randomness

Their Importance in the Complexity of Physical, Biological and Social Sciences

Unstable Singularities and Randomness

Their Importance in the Complexity of Physical, Biological and Social Sciences

By

Joseph P. Zbilut
Department of Molecular Biophysics and Physiology
Rush University Medical Center
Chicago, Illinois, USA

2004

ELSEVIER

Amsterdam – Boston – Heidelberg – London – New York – Oxford
Paris – San Diego – San Francisco – Singapore – Sydney – Tokyo

ELSEVIER B.V.	ELSEVIER Inc.	ELSEVIER Ltd	ELSEVIER Ltd
Sara Burgerhartstraat 25	525 B Street, Suite 1900	The Boulevard, Langford Lane	84 Theobalds Road
P.O. Box 211, 1000 AE Amsterdam	San Diego, CA 92101-4495	Kidlington, Oxford OX5 1GB	London WC1X 8RR
The Netherlands	USA	UK	UK

© 2004 Elsevier B.V. All rights reserved.

This work is protected under copyright by Elsevier B.V., and the following terms and conditions apply to its use:

Photocopying
Single photocopies of single chapters may be made for personal use as allowed by national copyright laws. Permission of the Publisher and payment of a fee is required for all other photocopying, including multiple or systematic copying, copying for advertising or promotional purposes, resale, and all forms of document delivery. Special rates are available for educational institutions that wish to make photocopies for non-profit educational classroom use.

Permissions may be sought directly from Elsevier's Rights Department in Oxford, UK: phone (+44) 1865 843830, fax (+44) 1865 853333, e-mail: permissions@elsevier.com. Requests may also be completed on-line via the Elsevier homepage (http://www.elsevier.com/locate/permissions).

In the USA, users may clear permissions and make payments through the Copyright Clearance Center, Inc., 222 Rosewood Drive, Danvers, MA 01923, USA; phone: (+1) (978) 7508400, fax: (+1) (978) 7504744, and in the UK through the Copyright Licensing Agency Rapid Clearance Service (CLARCS), 90 Tottenham Court Road, London W1P 0LP, UK; phone: (+44) 20 7631 5555; fax: (+44) 20 7631 5500. Other countries may have a local reprographic rights agency for payments.

Derivative Works
Tables of contents may be reproduced for internal circulation, but permission of the Publisher is required for external resale or distribution of such material. Permission of the Publisher is required for all other derivative works, including compilations and translations.

Electronic Storage or Usage
Permission of the Publisher is required to store or use electronically any material contained in this work, including any chapter or part of a chapter.

Except as outlined above, no part of this work may be reproduced, stored in a retrieval system or transmitted in any form or by any means, electronic, mechanical, photocopying, recording or otherwise, without prior written permission of the Publisher.
Address permissions requests to: Elsevier's Rights Department, at the fax and e-mail addresses noted above.

Notice
No responsibility is assumed by the Publisher for any injury and/or damage to persons or property as a matter of products liability, negligence or otherwise, or from any use or operation of any methods, products, instructions or ideas contained in the material herein. Because of rapid advances in the medical sciences, in particular, independent verification of diagnoses and drug dosages should be made.

First edition 2004

Library of Congress Cataloging in Publication Data
A catalog record is available from the Library of Congress.

British Library Cataloguing in Publication Data
A catalogue record is available from the British Library.

ISBN: 0-444-51613-1

∞ The paper used in this publication meets the requirements of ANSI/NISO Z39.48-1992 (Permanence of Paper).
Printed in The Netherlands.

Dedication

To Barbara

Preface

The present effort is an attempt to introduce ideas in more informal and general terms originally developed in an earlier work that was co-authored with M. Zak and R. Meyers (1997). In that work, a relatively detailed mathematical exposition regarding our theory of dynamic instabilities was introduced. Unfortunately, scientists having less formal mathematical training, or scientists outside the sphere of physical science were at a disadvantage. Nonetheless, many did plough through the work and found it provocative and useful for their own research. At the outset, it should be stressed that this book is not about the topic of randomness in general, complexity, chaos, stochastic resonance or similar derivative ideas (although there is not doubt that many of the ideas here are a result of advances in these areas). Rather, it is a presentation of a novel paradigm for understanding dynamics as viewed by several formal disciplines. In contrast to the many geometric explications of dynamics, the emphasis here is probability; i.e., the statistical aspects of the unique motions that are punctuated by certain singularities.[1]

In October of 2002, a general introduction to this author's theory of instability and singularities was presented as part of the Master Lecture Series on the topic of Complexity of Neuroscience, of the Institute of Psychiatry and Clinical Psychology of the Catholic University of the Sacred Heart, Rome, under the direction of Professor Sergio De Risio. At that time, Professor De Risio urged the author to develop a general introduction to these ideas suitable for a wide range of scientists, researchers, and students. This monograph is the result.

---

[1] There is a formal mathematical (and physical) discipline of singularities, which I refer to at times. However, this work is not intended in the same sense.

It was originally intended to be brief, but requests from different sources have expanded it beyond its original intention. From the outset, the idea has been to present the main conceptual paradigms for these singularities, so that they may be accessible to a wide variety of investigators across several disciplines. At the same time, some effort has been made to clearly distinguish these lines of thought from other concepts popularized in nonlinear dynamics.

An attempt has been made to avoid mathematical formalisms. Initially, I had intended to avoid mathematics altogether; however, several reviewers requested the inclusions of examples to help clarify points. Those that do remain have been chosen mainly to specify and illuminate certain ideas, and should not be beyond the abilities of the average scientist. Examples have been chosen which are as "basic" as possible, as well as varied as is possible. Often, in this respect I am indebted to scientists in other fields—whom I list below. Also, specific details of analysis and methods have also been largely avoided. For those interested, the previously mentioned work (1997) presents many of the mathematical details. A glossary has been added to further clarify terms, as well as a mathematical appendix which goes into some practical details involved not only in the theory of nondeterministic dynamics, but also into some practical methods to identify them in real data. It should be also noted that the ideas presented here are specifically the author's and not necessarily coincident with those of colleagues he has worked with. Although I am sure they would agree with many of the points, it would be unfair to assume complete agreement.

Although I had not intended originally to digress into philosophical issues regarding science, epistemology and ontology, I found myself doing so in order to complete the discourse. In discussing some of these issues with colleagues, I realized I could not avoid the responsibility of framing the derivative issues. As with many of the topics I bring up, I leave to others more schooled in specific areas to advance the discussion.

Certainly, no scientist works in a vacuum, and I am indebted to my friends who have provided the requisite "noise" and motivation for these ideas. Specifically, in alphabetical order: Rita Ballochi, Consiglio Nazionale Ricerche Istituto di Fisiologia Clinica, Pisa, for her continued interest in many of these ideas; Alfredo Colosimo, University of Rome "La Sapienza", for his constant push to refine; Elio Conte, University of Bari, for his new insights into the paradigm and compartment models; Filippo Conti, University of Rome "La Sapienza", for his ability to put together many disparate ideas; Dave Dixon, University of California, Riverside, for his incisive but studied appreciation of singularities; Antonio Federici of the University of Bari, for his grasp of the idea into new areas; Sandro Giuliani, Istituto Superiore di Sanitá, for his simple correlations; Cesare Manetti, University "La Sapienza", for his understanding of molecular motions; Paul Rapp, Drexel University College of Medicine, for his gentlemanly approach to a stochastic world; Chuck Webber, Loyola University Medical Center, Maywood, IL, for his constancy; Gene Yates, University of California, Los Angeles, for his appreciation of the dynamics of life; and Mike Zak, Jet Propulsion Laboratory, California Institute of Technology, for his encyclopedic knowledge of mechanics.

I would also like to thank Tom Hoeppner of the Neurology Department of Rush Medical College, and Nitza Thomasson, formerly my post-doctoral fellow thanks to INSERM, Paris, for their work on EEGs; Josema Comenges-Zaldivar at the European Commission Research Center at Ispra, and Fernanda Strozzi and Jordi Bosch, Carlo Cattaneo University at Castellanza for their collaboration with chemical reactions as well as financial series; and Maurizio Tomasi for pointing out some interesting aspects pertaining to the Spanish Steps. I am also indebted to Franco Orsucci of the Institute for Complexity Studies in Rome for providing me the opportunity to present these ideas in the first place.

Portions of this work were supported by a grant from the National Science Foundation, Division of Mathematical Sciences, #0240230 supporting work in mathematical biology. I thank Springer-Verlag for allowing me to use graphics from the earlier work with Mike Zak and Ron Myers; these are identified as "ZZM" in the text. Appreciation is expressed to the editors of *Journal of Physics A*, for giving permission to reproduce figures in Dave Dixon's article—and to Dave Dixon himself for presenting some of our early combined works and his exposition on the simple harmonic oscillator and magnetic field equations. Thanks are due for the continued support of Bob Eisenberg, Kay Andreoli, Margaret Faut, and Kathy Lauer of Rush University Medical Center. My appreciation is also expressed to Ms. Jeanette Bakker of Elsevier for making the entire process of developing this book a pleasure. Finally, I thank Sandro Giuliani for reading a preliminary version of this work—se sono rose, fioriranno. As usual, his comments were illuminating.

I have tried to maintain a decided informality in these pages. I along with some of my colleagues have in large measure published more formal expositions of some of these ideas, many of which are included in the bibliography. The intent here was to keep the spirit which Professor De Risio originally suggested—something which would not intimidate students from as diverse a population as possible.

In keeping with this spirit the organization of the book has had as its objective the presentation of the material in its most basic form. For those, however, wishing more specifics I have provided material generally under third level headings, which may be skipped initially. For those looking for specific techniques useful for analyzing data, material has been provided in the Mathematical Appendix.

Contents

PREFACE VII

1 PROBABILITY AND DYNAMICS 1

1.1 A Dichotomy 1

1.2 Historical Perspective 2

1.3 Probabilities 3

1.4 Randomness 5

1.5 Singularities 5

1.6 Models and Reality 7

2 SINGULARITIES AND INSTABILITY 11

2.1 Dynamics 12
  2.1.1 Attractors 13
  2.1.2 Liapunov Exponents 15

2.2 Limitations of the Classical Approach 16

2.3 Dynamical Instability 18

2.4 Lipschitz Conditions 21

2.5 Basic Concepts 25
    2.5.1 Dissipation 26
    2.5.2 Terminal Dynamics Limit Sets 28
    2.5.3 Interpretation of Terminal Attractors 29
    2.5.4 Unpredictability in Terminal Dynamics 30
    2.5.5 Irreversibility of Terminal Dynamics 30
    2.5.6 Probabilistic Structure 31
    2.5.7 Self-Organization in Terminal Dynamics 32

3 NOISE AND DETERMINISM 35

3.1 Experimental Determinations 42

3.2 The Larger Metaphor 45

3.3 Non-Equilibrium Singularities 50
    3.3.1 Simple Harmonic Oscillator 50
    3.3.2 A Physically Motivated Example 53
    3.3.3 Uncertainty in Piecewise Deterministic Dynamics 57
    3.3.4 Nondeterminism and Predictability 62
    3.3.5 Controlling Nondeterministic Chaos 64
    3.3.6 Implications 67

3.4 Classification of Nondeterministic Systems 68

4 SINGULARITIES IN BIOLOGICAL SCIENCES 73

4.1 An Alternative Approach 75

4.2 Nonstationary Features of the Cardio-Pulmonary System 79
    4.2.1 Tracheal Pressures 81
    4.2.2 Lung Sounds 81
    4.2.3 Heart Beat 84

4.3 Neural (Brain) Processes 89
    4.3.1 Electroencephalograms and Seizures 90
    4.3.2 Terminal Neurodynamics 98

- 4.3.3 Creativity and Neurodynamics 101
- 4.3.4 Collective Brain 102
- 4.3.5 Stochastic Attractor as a Tool for Generalization 103
- 4.3.6 Collective Brain Paradigm 106
- 4.3.7 Model of Collective Brain 107
- 4.3.8 Terminal Comments 108

4.4 Arm Motion 109

4.5 Protein Folding 112
- 4.5.1 Two General Contemporary Schemata 113
- 4.5.2 A Different View 114
- 4.5.3 Proteins from a Signal Analysis Perspective 115
- 4.5.4 Singularities of Protein Hydrophobicity 116

4.6 Compartment Models 126

4.7 Biological Complexity 135

5 SINGULARITIES IN SOCIAL SCIENCE/ARTS 143

5.1 Economic Time Series 144
- 5.1.1 Stock Market Indexes 145
- 5.1.2 Exchange Rates 149

5.2 Art (The Science of Art?) 149
- 5.2.1 Examples 151

5.3 Psychology 158
- 5.3.1 Examples 158

5.4 Sociology 160
- 5.4.1 Examples 161

6 CONCLUSIONS 165

7 GLOSSARY 169

8 MATHEMATICAL APPENDIX 181

8.1 Nondeterministic System with Singularities 181

8.2 Recurrence Quantification Analysis (RQA) 183

8.3 Recurrence Plots 184

8.4 Recurrence Quantification 190
    8.4.1 Determining Parameters for Nonstationary Series 195
    8.4.2 Choice of Embedding 195
    8.4.3 Choice of Lag 195
    8.4.4 Choice of Radius 196

8.5 Detecting Singularities 197
    8.5.1 Maxline (Liapunov exponent) 198
    Real Signal 198
    Results 198
    Batch Reactions 204
    Experiments 208
    Results 210
    8.5.2 Orthogonal Vectors 212
    8.5.3 Some Observations 213

BIBLIOGRAPHY 221

Books 221

Book Chapters, Proceedings 223

Journals 224

INDEX 231

1 *Probability and Dynamics*

> *I will call myself the child of chance.*
>
> —*Oedipus*
>
> The Oedipus Tyrannos of Sophocles.
> Boston, Ginn Brothers, 1874.

1.1   A Dichotomy

The notion of complexity as an object of scientific interest is relatively new. Prior to the 20th century, the main concern was that of simplicity, with the implication that there existed an opposite pole of complexity. Thus there existed the insinuation that scientific epistemological perception was defined as a delicate balancing act. In some sense, this idea is anomalous to the whole modern scientific enterprise as developed from the 16th century onward: the simplicity/complexity dichotomy is essentially a product of Medieval Scholastic philosophy which was rooted in deductive reasoning based on first principles. The modern scientific revolution, on the other hand was developed on the idea of inductive, observation-dependent reasoning. Thus "reason" is not really rejected, but transformed as an intermingling of deduction and induction. Thus, for some three centuries, the scientific enterprise has been based on varying degrees of a priori conceptualizations combined with real observations and testing. A dominant motive behind these outward manifestations of science ultimately still redounded to a search for "causality." What was essentially a noetic endeavor for scholastics, became an observational puzzle for scientists. The idea that the universe was driven as some wondrous machine could not be abandoned. The only role for probability was left to the world of games and "chance."

1.2   Historical Perspective

Simplicity, as a formal principle, has had a long history enshrined in the dictum of the 14th century Franciscan philosopher, William of Occam, who postulated that "pluritas non est ponenda sine necessitate" [being is not multiplied without necessity], and passed on simply as "Occam's razor," or more prosaically, "keep it simple."

Indeed, the razor has been invoked by such notables as Isaac Newton, and in modern times, Albert Einstein and Stephen Hawking to justify parsimony in the adoption of physical principles. Although the dictum, has proved its usefulness as a support for many scientific theories, the last century witnessed a gradual concern for simplicity's complement. Implicit was a recognition that beyond logical partitions, was a need to quantify the simple/complex continuum. Perhaps the first milestone in the road to quantifying complexity came with Claude Shannon's famous information entropy in the late 1940's. Although it was not specifically developed as a complexity measure, it soon provided an impetus for sustained interest in information as a unifying concept for complexity. What was remarkable about Shannon's work was the fact that he used stochastic methods to "quantify" message content. This is to point out that statistics, as opposed to formal, logical mathematics, was used.

Although most scientists lump mathematics as one monolithic subject which may be useful to their own activities, the stochastic/formal dichotomy essentially supports the simple/complex paradigm, as well as the deductive/inductive and first principles/observation approaches: formal, logical mathematics is based on an assumption of elegant determinism, whereas stochastics look to probabilities. To the present day, the scientific endeavor is variously described as an attempt to understand the world in terms of patterns or structures in an ensemble of events. Most frequently, this involves the elaboration of deterministic laws which could be described by systems of differential equations. Yet, as physical scientists of the 19th century became focused on the microscopic scales of phenomena, as opposed to the macroscopic scales, they encountered increasing levels of "noise." To deal with this noise, mathematical statistics was entrained to provide predictive quantification. Subsequently, the kinetic theory of gases and quantum mechanics came into being, while determinism became a limiting case of statistical formulations. Thus, we are faced with a curious scenario: inductive reasoning and its generated statistical ensembles are simultaneously convolved with the attempts at deterministic prediction of formal, logical mathematics. Inherent in such efforts are inconsistencies. As David Mumford has pointed out, mathematics is basically a project of reproducible mental objects associated with rules. Consequently, there is a constant interplay between observation with mental concepts: Hume and Bacon have never been resolved. Instead, a pragmatic accommodation has been in place for centuries. Induction and deduction have informed each other, and if it "works," so be it. What should be acknowledged is that stochastic formulations of observed phenomena deserve a greater role in scientific endeavors: the reductionist attitude in modern science, although useful, has fostered an assumption that every phenomenon is unique—with variability being primarily dependent upon experimental vagaries. This has been especially true in the biological sciences. On the contrary, phenomena may not be rigidly deterministic: the remarkably successful history of life as we know it may be due to its stochastic flexibility.

1.3 *Probabilities*

That it took until the middle of the last century for the notion of probability to develop is in some sense remarkable. The beginnings of the 20th century witnessed a curious development in the history of science. The applications of statistics to the diverse phenomena of the biological and social sciences were about to explode as a result of the work of such people as Pearson and Fisher. On the other hand, the world of physical sciences was still avoiding the use of stochastic models, although Boltzmann and Gibbs had supplied sufficient reason not to do so. Now, at the beginning of the 21st century, the state of affairs has changed considerably. The physical sciences have come to appreciate the significant insights into low dimensional systems which appear speciouslyrandom, while at the same time, the biological and social sciences are increasingly interested in deterministic descriptions of what appear to be very complex phenomena. The intersection of these two approaches would appear to be the often neglected, sometimes unwanted phenomenon of "noise." Often relegated to status as nuisance, noise has become more appreciated really as "that which we cannot explain." And in this explanation noise has become recognized as a possible deterministic system[2] itself, perhaps involved with quantum effects, with a complicated description that interacts with observables on a variety of length scales. At the juncture between the physical and biological this noise creates myriad effects which ultimately redound to the very basic ideas regarding the constitution of what we know as living matter.

That this should be so is not surprising. Although it was not that long ago that scientists felt that a understanding of classical Newtonian laws of motion could provide the key to understanding all of existence, the current climate appreciates that with some modifications, this might still be true: the movements of ions through cellular channels are being investigated by biologists with a seriousness that would be the envy of an experimental physicist. Indeed some of the very time-honored models of the physical sciences such as spin lattices are being used to explore this area. And why not? At this level the very fundamental laws of physics control discrete molecular events which have profound importance for living tissues. Ultimately, the dynamics at this level govern the way neurons, and other humoral agents which orchestrate the myriad events to maintain the human organism. "Neural nets" are once again studied as true models of the nervous system, not only by biologists, but by physicists as well.

The flurry of activity in this broad area is not unremarkable given that biological systems are often poorly defined. Until the present, most of our understanding of biological systems has been delimited by phenomenological descriptions guided by statistical results. Linear models with little consideration of underlying specifics have tended to inform such processes. What is more frustrating has been the failure of such models to explain transitional, and apparently aperiodic changes of observed records. The resurgence of nonlinear dynamics has provided an opportunity to explain these processes more systematically, and with a formal explanation of transitional phenomena.

Certainly, nonlinear dynamics is not a panacea. Linear descriptions do, in fact, account for many biological and social processes. Additionally, there is the danger to assume that chaotic correspondence with experimental data "explains" the system. Scientists are all too familiar with the pitfalls of model-making. Mathematics is the language of science, but the language is not the science. And the underlying idealist procedures of mathematics can provide sometimes true and sometimes false descriptions of phenomena. Physics itself is replete with

---

[2] Although the term, "system" is used freely, especially in the context of biology, it may not be quite appropriate given its common usage in mechanics. The reasons for this are made more explicit later, but briefly, it gives the idea that organisms are well designed rational machines, which may not be the case.

examples of this tension between mathematics and reality. Consider for example the debates regarding delta functions, and "infinitesimals." It was Einstein himself who cautioned about the interface between mathematics and the physical sciences.

1.4 *Randomness*

At the same time there is the ever present concern that by learning about the intricacies of the processes, we neglect the global kinetics of a system. As Laughlin, et al. have remarked:

"Although behavior of atoms and small molecules can be predicted with reasonable accuracy starting from the underlying laws of quantum mechanics, the behavior of large ones cannot, for the errors always eventually run out of control as the number of atoms increases because of exponentially increasing computer requirements. At the same time, however, very large aggregations of particles have some astonishing properties, such as the ability to levitate magnets when they are cooled to cryogenic temperatures, that are commonly acknowledged to be 'understood.' How can this be? The answer is that these properties are actually caused by collective organizing principles that formally grow out of the microscopic rules but are in a real sense independent of them."

Continuing evidence suggests that there is a constant interplay between microscopic and macroscopic length scales, as well as randomness to create enormous variety and patterns in biology. And perhaps this is the important point that has emerged: we have traditionally maintained a perspective of looking for order, and disregarding randomness and instability as a nuisance; whereas the correct perspective may be to see this nuisance as an active process which informs order and vice versa.

1.5 *Singularities*

The natural extension of this view is to ask how this "nuisance" exists and works? The perspective taken here is to attempt to understand biological and related systems (personal, societal) in a unique way, and this unique way involves the admittance of singularities both mathematically and biologically that are at the heart of the nuisance. In this endeavor I refer to the comments made by James Clerk Maxwell over a century ago when he pointed out:

"Every existence above a certain rank has its singular points: the higher the rank the more of them. At these points, influences whose physical magnitude is too small to be taken account of by a finite being, may produce results of the greatest importance. All great results produced by human endeavor depend on taking advantage of these singular states when they occur."

And it is the singularity which is often a troublesome prospect for mathematical thinking. Some have called them pathologic—to be done away with.

Certainly, biological organisms are of a high rank, and indeed, many of these singularities have already been uncovered. From a topological perspective, Winfree has demonstrated time and again that biological oscillators admit singularities. Other work has argued from first principles and experimentation that physiological singularities must exist in order for the organisms to maintain adaptability. What has not been adequately appreciated is the reconciliation between classical Newtonian dynamics and these biological/social phenomena. This work represents a modest attempt in this direction. In order to proceed, certain problems in classical dynamics need to be highlighted.

Classical dynamics describes processes in which the future can be derived from the past, and past can be traced from future by time inversion. The implications of such a formulation are rarely considered since it usually works so well. But because of this determinism, classical dynamics becomes fully predictable, and therefore it cannot explain the emergence of new dynamical patterns in nature, in biological, and in social systems. This major flaw in classical dynamics has attracted the attention of many outstanding scientists (Gibbs, Planck, Prigogine, etc.). Recent progress in understanding the phenomenology of nonlinear dynamical systems was stressed by the discovery and intensive studies of "chaos" which, in addition to a fundamental theoretical impact, has become a useful tool for several applied methodologies. However, the, actual theory of chaos has raised more questions than answers. Indeed, how fully deterministic dynamical equations with small uncertainties in initial conditions can produce random solutions with a stable probabilistic structure? And how can this structure be predicted? What role does chaos play in information processing performed by biological systems? Does it contribute into elements of creativity, or irrationality (or both!) in the activity of a human brain? In the past I have argued that these chaotic systems cannot be the basis of many real systems for one major reason; namely, the burden of their past: they maintain a huge memory which adds to their inertia, and prevents them from being able to make necessary adaptations quickly and with a minimum of effort. This is a subtle point that is easily overlooked when admiring the "predictability" of a given deterministic solution to a phenomenon. Moreover, since predictability is a hallmark of good science, to put this endeavor into question, is to risk scorn. This is not my aim; instead, the aim is to question the appropriate form of the predictability. Certainly, as has been pointed out, quantum physics must content itself with probabilistic solutions to many of its problems. These questions have motivated this study, and the answers center around the concepts of discrete events, singularities, their instability and stochasticity.

1.6 Models and Reality

An important distinction in these discussions is the fact that mathematical and experimental (real) models do not necessarily coincide—a fact well appreciated by Einstein. This is not to say that the mathematics is wrong per se, but that mathematics is based on its own logical set of axioms—which do not always parallel reality. Insofar this is true, a mathematical model is an approximation. The difficulty arises when mathematical models appear to predict the reality. Nowhere has this been more adequately summarized than in the experience with the Ptolemaic vision of the earth being at the center of the universe: the model could easily predict planetary motions. Nonetheless, it was wrong, and it was centuries before it was challenged by Copernicus. In this sense, many currently viewed ideas, although not necessarily completely wrong, suffer from small problems which can "blow up." Nonlinear dynamics has shown that these "small problems" can have a massive long-term effect on the evolution of a process. It is curious that this type of problem has been in existence for many centuries, and perhaps parallels our own concerns with the way reality and its apprehension by the mind works.

A useful example is "Buridan's ass." Jean Buridan was a 14th century French philosopher, who interestingly, studied under William of Occam. He is noted to have posited the problem of an ass which was both hungry and thirsty, and was placed at an equal distance between food and water—what would it choose first? Or would the ass be so confused, that it would do nothing and die. Although this anecdote has been attributed to him, it has not been found in any of his writings. Indeed, various versions of this story have been traced as far back as Aristotle. The point to be made, however, is that the decision constitutes a singularity—a unique situation. Mathematically, it may be considered as a metastable state defined by a singularity—something encountered in asynchronous computer functions, and has recommended a variety of solutions (Fig. 1.1).

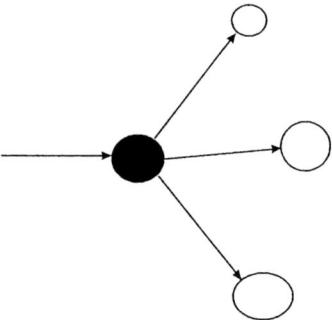

Figure 1.1. A singularity is a peculiar point (in black) in a dynamics which has no real connection to the past and may have multiple responses. Movement away from the singularity may be initiated by an infinitesimal amount of noise, or more structured forces. In real life movement always occurs because there is always some "noise."

This book provides a different perspective in the study of nonlinear dynamics with additional attention to biological and social systems. The main idea is that although there have been many useful advances in these dynamics through "chaos" theory, this theory is for deterministic systems and thus not appropriate for the description of biological and other systems. Whereas "chaos" is sensitive to initial conditions, it is globally "stable." In other words, generally, changing initial conditions can result in different trajectories, but its attractor basin is unchanged. As a result, perturbations do not change the global dynamics. On the other hand, it has been observed that "noise" plays an important role in many systems. This suggests that many complex systems not only may have "global stability," but also "local instability" usually caused by noise. The recommendation to take account of both local instability and global stability devolves to a theory of singularities and nondeterministic dynamics. At singular points a small noise imposes a dramatic influence on the system's behavior, while it can be neglected in non-singular regions.

An alternate view is that unpredictability can be based on singular dynamics. The main emphasis is on intrinsic stochasticity caused by the instability of the dynamics and their equations. This approach is based upon a revision of a mathematical formalism of Newtonian dynamics, and, in particular, upon elimination of requirements concerning differentiability, which is some cases can lead to unrealistic solutions. Although apparently revolutionary, this is not the case: it is well appreciated that the underlying concepts of reality are quantum, and that Newtonian dynamics are an approximation dependent upon scales of observation. Yet even in the usual human scale it is argued that there are deficiencies. These deficiencies are especially important when dealing with the biological or social; i.e., where physical "precision" gives way to probabalism. And it is probabalism, perhaps, that is the reason for human creativity—and sorrow, as Oedipus asserts.

This revision allows us to reevaluate our view on the origin of "chaotic" dynamics, on prediction of their probabilistic structures, and on their role in biological and social systems. For the treatment of the material, insight is emphasized rather than mathematical rigor with few exceptions. The approach taken is essentially conceptual and historical from the view of my own experience with these problems.

For many analyses, I depend upon recurrence quantification analysis (RQA)—a method which appears to be uniquely useful in uncovering singularities. The Mathematical Appendix contains an enlarged exposition of the method and its applications.

2 *Singularities and Instability*

Before we can talk about singularities, a brief introduction into the area of dynamics is in order. This introduction is not intended to be comprehensive, since the territory has been covered by numerous books, especially in recent history. By now many readers have been exposed to terms such as "attractors," "bifurcations," etc. As with any time series analysis (and in most cases our appreciation of dynamics is in time, or at the very least some ordered series), however, significance can often be found in small details, perhaps phasic as opposed to level effects. Certainly, the thesis presented here is dependent upon an appreciation of such small details. There is a qualitative difference between linear and nonlinear series, most notably in their appreciative phenomenology. Consequently, this presentation is biased with a view to explaining the notion of singular dynamics.

Traditionally, dynamics is presented as part of classical physical mechanics. However, this traditional presentation has assumed an underlying determinism: Newton's laws and derivative methods such as Hamiltonians and Lagrangians have dominated such a discussion in attempts to accommodate the deterministic aims. I will avoid these points, since these have been technically addressed by our previous work. Instead, I will emphasize the significant differences implied in the "nondeterministic" singular point of view. As such I will avoid a "canonical" view, mainly from the desire to avoid preconceptions, as well as from a desire to create some "dissonance" in commonly held ideas. Chief among these ideas are those of causality, determinism, stability. Often they possess an underlying ethical connotation of goodness, with their opposites being "bad." It is perhaps a human trait to wish to moralize phenomena—expressing more about the consequences for the person than for the phenomena. Nonetheless, this moralization of observable phenomena can needlessly complicate a dispassionate discourse by "structuring" experience before it happens.

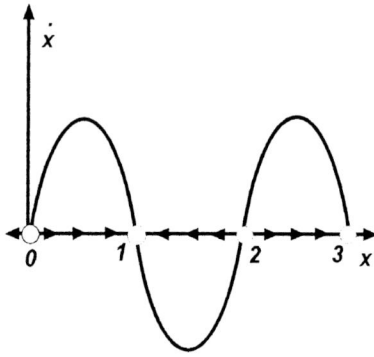

Figure 2.1. Static attractor and repeller (ZZM).

2.1 Dynamics

Dynamics describes the motion of systems, i.e., the time evolution of its parameters. The time variable can be discrete or continuous. In discrete-time dynamical systems, the rate of change of their parameters is defined only for discrete values. These systems can be presented as the iteration of a function; i.e., as difference equations.

In continuous-time dynamical systems the rate of change is defined for all values of t; such systems can be modeled by ordinary differential equations, or by partial differential equations.

$$\frac{dx}{dt} = \dot{x} = \upsilon(x,t) \qquad (2.1)$$

or by partial differential equations:

$$\dot{x} = \upsilon(x, x', x'', \ldots, t), \; x' = \frac{\partial x}{\partial s}, \; x'' = \frac{\partial^2 x}{\partial s^2}, \ldots \qquad (2.2)$$

if the rate of change, in addition, depends upon distributions of x over space coordinates, s. In these equations x represents the state of the dynamical system.

Continuous-time dynamical system theory has adopted the basic mathematical assumptions of the theory of differential equations such as differentiability of the parameters (with respect to time and space), the boundedness of the velocity gradients (termed Lipschitz conditions), etc. Under these assumptions, the existence, uniqueness and stability of solutions describing the behavior of dynamical systems has been studied. In short, physicists and mathematicians have elaborated significant models describing dynamics. One consideration, however, is that many of these models have limited practicability when applied to living systems. Specifically, living systems are not idealized, stationary systems. They are most often dissipative; i.e., they tend to be time irreversible with friction—a feature which has been the object of life-long study for the Nobel laureate, Ilya Prigogine. Nonetheless, some of the concepts are useful if the appropriate limitations are considered.

2.1.1 Attractors

One of the most commonly used terms in describing dynamics is that of attractors (see Glossary). For the case of Eq. 2.1,

$$\dot{x} = \sin x \qquad (2.3)$$

and find equilibrium points, i.e., points with zero velocity:

$$\dot{x} = \sin x_k = 0, \frac{\pi k}{2}, k = \cdots -2, -1, 0, 1, \cdots \quad (2.4)$$

the equilibrium points are stable, and are called static attractors, "attracting" all nearby points to itself. So for example, if the "system" were originally located at an attractor, say attractor 1, it will remain there forever. If the system started from a position between points 0 and 2 (a "basin" of attraction), it will eventually approach the attractor 1 during an infinitely large period of time (Fig. 2.1)! Consequently, a dynamical system in an open interval between two static attractors cannot pass either of them, and cannot escape the basin of attraction. The basins of attractors are separated by static repellers—the unstable equilibrium points $\cdots 2, 0, 2, 4, \cdots$.

Static attractors and repellers, in general, can be found in autonomous dynamical systems, i.e., systems which are not subjected to any external influences that depend on time, so that

$$\frac{\partial v}{\partial t} = 0 \qquad (2.5)$$

Attractors can also be periodic (Fig. 2.2):

$$\dot{r} = -\sin r, \dot{\theta} = \omega = \text{Const} \qquad (2.6)$$

where r and θ are polar coordinates.

The same attractors are available for r as x in the previous example. In this case, however, the attractors are not static, since the total velocity of the system is not zero, and instead the system approaches states which are characterized by periodic motions:

$$\theta = \theta_0 + \omega t \qquad (2.7)$$

with a period of $2\pi/\omega$. These states are called periodic attractors, and are stable with respect to r. The coordinate θ, however may be at the boundary of stability, and small errors will increase in it (but in a linear—not exponential fashion.

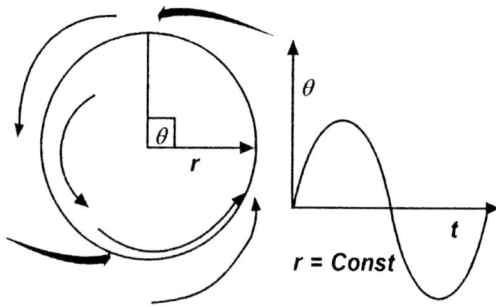

Figure 2.2. Periodic attractor, and its time series.

2.1.2 Liapunov Exponents

Perhaps one of the most useful, and frequently used concepts in dynamics is that of Liapunov exponents. It is the averaged rate of divergence (or convergence) of two neighboring trajectories (Fig. 2.3).

$$\sigma = \lim_{\substack{t \to \infty \\ d(0) \to 0}} \left(\frac{1}{t}\right) \ln\left(\frac{d(t)}{d(0)}\right) \quad (2.8)$$

where $d(0)$ and $d(t)$ initial and current distances, between two neighboring trajectories in certain directions. Unfortunately, one of the problems is that Liapunov exponents cannot, in general, be analytically expressed via the parameters of the dynamical system, making the idea of prediction difficult. (One can get an approximation, however, using a feature of recurrence analysis. See the Mathematical Appendix.)

To be sure, a significant contemporary use of these exponents has been to demonstrate the existence of chaotic dynamics; i.e., sensitive dependence on initial conditions. As a practical matter, in the case of many experimental systems, confirmation of chaos is doubtful, given the data and stationarity needs for the unambiguous calculation of Liapunov exponents. This is to emphasize the fact that it is one thing to calculate such values for clean computer models of chaotic dynamics; whereas real systems—especially biological—are inherently noisy, and non-stationary.

Certainly, the need to confirm chaos is often questionable, as Ruelle as pointed out. To claim that a dynamic is deterministically "chaotic" requires considerable effort. Use of chaotic invariants such as dimensions, entropies and Liapunov exponents are notoriously suspect given the stringent demands for their application. Additionally, one can question the utility of such an appellation—is it any more significant than saying some process exhibits a Gaussian distribution?

Nonetheless, if considered within the context of their development, positive Liapunov exponents can provide important information apart from notions of chaoticity. Liapunov exponents were developed to characterize a system's stability/instability, which of itself may be an important finding.

The dissipativity of a dynamical system does not exclude the possibility that some of the Liapunov exponents are positive. Attractors can be static and periodic (Figs. 2.2, 2.3). Thus, chaotic attraction emerges from a coexistence of two conflicting phenomena: dissipation which contracts the volume occupied by the motion trajectories, and instability which diverges these trajectories in certain directions. As a result of this, the trajectories are mixing, and a limit set, i.e. a chaotic attractor is developed. However, a big question can remain: is there an "attractor"? Crutchfield has pointed that theoretically, if the dynamics is transient, and if there is an attractor, it may take an extremely long time to determine if attraction is a possibility. The implication here is that the existence of an attractor is a "teleological," deterministic formulation implying that the dynamics are "slaved" by their history.

Obviously, a replacement of all the positive Liapunov exponents by zeros leads to periodic or multi-periodic attractors which represent the boundary between predictable and unpredictable motions.

2.2 Limitations of the Classical Approach

To evaluate the level of congruence between classical formulations of biological models and real systems, it may be useful to consider "neural nets." One of the most interesting and common paradigms for the joining of physical principles with biology is that of neural nets. Originally developed as an actual model for neuronal processing, they have become a popular method of information processing. The biggest promise of neural networks as computational tools lies in the idea that they resemble information processing in

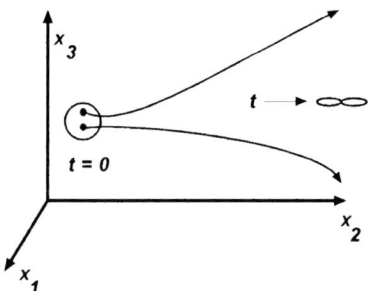

Figure 2.3. Liapunov exponents. Exponential divergence of nearby trajectories (ZZM). Note that the initial positions (in circle) are uncertain relative to the precision of the measuring instrument. In principle, if infinite precision were attainable, the trajectories could easily be predicted.

biological systems. What is often not recognized is that models based upon classical dynamical system theory are characterized by some limitations which prevent their acceptance with their real biological counterparts.

First, the neurons performance in this model is collective, but not parallel: any small change in the activity of a given neuron instantaneously effects all other neurons: in contrast to this, all biological systems exhibit both collective and parallel performances. For instance, the right and the left hands are mechanically independent (i.e., their performance is parallel), but at the same time their activity is coordinated by the brain; i.e., this makes their performance collective. This is a subtle but important distinction.

The performance of the model is fully determined by its initial conditions. The system never "forgets" these conditions: it carries this historical "burden" to "infinity." In order to change the system performance, the external input must overcome its "history." In contrast to this, biological systems are much more flexible: they can forget (if necessary) the past, adapting their behavior to environmental changes.

Finally, the features characterizing the system are of the same scale: they are insulated from the microworld by a large range of scales. At the same time, biological systems involve mechanisms that span the entire range from the atomic, to the molecular, to the tissue, etc. Presumably, real systems allow the different scales to communicate (in terms of information exchange) with each other.

It is questionable whether these limitations can be removed within the framework of classical dynamics. Certainly, all the systems considered in classical dynamics satisfy the Lipschitz condition which guarantees the uniqueness of the solutions subject to prescribed sets of initial conditions. For a neural system this condition requires that all the derivatives exist and are bounded:

The uniqueness of the solution subject to given initial conditions can be considered as a mathematical interpretation of rigid, predictable behavior of the corresponding dynamical system. Therefore, all the limitations of typical neural net models mentioned above are inevitable consequences of the Lipschitz condition and therefore of determinism of classical dynamics.

Classical dynamics describes processes in which the direction of time does not matter: its governing equations are invariant with respect to time inversion, in the sense that the time backward motion can be obtained from the governing equations by time inversion. Prigogine has stressed that in this view, future and past play the same role: nothing can appear in future which could not already exist in past since the trajectories followed by particles can never cross. This means that classical dynamics cannot explain the emergence of new dynamical patterns in nature in the same way in which nonequilibrium thermodynamics does. This is why the discovery of chaotic motions (which could lead to unpredictability in classical dynamics) originally disturbed the scientific community. Closer examination of such dynamics, however, demonstrate that they are still fully deterministic: the time of approaching attractors as well as the time of escaping repellers becomes theoretically infinite. They are also slaved by their history. A clear reason for such a paradigm is the requirement that their "equations" have unique solutions; i.e., they exhibit Lipschitz conditions. If this requirement is removed, the dynamics are, in a sense, liberated: the time to attractors and repellers are finite. Also, at repellers the solution becomes totally unpredictable within the deterministic mathematical framework, but it remains predictable in probabilistic sense. In contrast to "classical" chaos here the randomness is generated by the differential operator itself as a result of the failure of uniqueness conditions at the equilibrium points.

2.3 Dynamical Instability

The word "instability" in common language has a connotation of something going wrong, being unreliable, or being abnormal: the rapid growth of cancer cells, "irrational" behavior of a patient, collapse of a structure, etc. However, instabilities can be beneficial, which lead to positive change, evolution, progress, and creativity: they can explain nondeterministic (multi-choice) behavior in physical, biological and social systems—as well as the science of art.

The concept of instability is an attribute of dynamical models which describe change in time of physical parameters, biological or social events, etc. Each dynamical model has a certain sensitivity to small changes or "errors" in initial values of its variables. These errors may grow in time, and if such growth is of an exponential rate, the behavior of the variable is defined as unstable (Fig. 2.4). However, the over-all effect of an unstable variable upon the dynamical system is not necessarily a destructive process.

When dynamical models simulate biological, or social behavior, they should include the concept of "discrete events," i.e., special critical states which give rise to branching solutions, or to bifurcations. Obviously, the condition of uniqueness of a solution at the critical points must be relaxed. Thus, driven by alternating stability and instability effects, such systems perform a "statistical" behavior whose complexity can match the complexity of biological and social worlds.

Most dynamical processes are so complex that a universal theory which would capture all the details during all time periods is unthinkable. (As an example, consider the documented difficulty of developing reliable models of atmospheric dynamics suitable for weather prediction. Currently, there are several models running on powerful computers—all disagreeing with each other to some degree. Financial markets have spent tremendous sums in hopes of developing models which could aid in predicting trends—and yet one reads periodically that random choice of stocks often works as well as any model. And in the biological realm, reliable models of heart functioning have been extremely difficult to produce. That is why the art of mathematical modeling is to extract only the fundamental aspects of the process and to neglect its insignificant features, without losing the core of information. But the determination of insignificant features is not a simple. In many cases even apparently trivial small forces can cause large changes in dynamical systems. Such circumstances quite naturally result in instability. Obviously such forces cannot be considered as "insignificant," and therefore, they cannot be ignored. Often, these are a result of "nonlinearities," i.e., cause and effect situations which are not "proportional." Engineers appreciate these nonlinearities extremely well—failure to detect or understand could compromise public health and safety. Unfortunately, at their inception, it is difficult to incorporate them into a model. This simply means that the model is not adequate for a quantitative description of the corresponding dynamical process--it must be changed or modified. However, the instability delivers important qualitative information—it manifests the boundaries of applicability of the original model.

Instabilities can be relatively short or long-term. Short-term instability occurs when the system can end up in stable states. For dissipative systems such states can be represented by so-called static or periodic "attractors," i.e., a state to which the system converges. In the very beginning of the post-instability transition period, the unstable motion cannot be traced quantitatively, but it becomes more and more deterministic as it approaches the attractor. Hence, a short-term instability does not necessarily require a model modification. It is sometimes considered a "transient." Long term instability occurs when the system does not have an alternative stable state. The difficulty, however, is to determine when a transient is finished; i.e., is it long or short? Crutchfield, again has suggested that in some cases one could never practically know not only that the dynamics are on a transient, but also when they will end. (It has been quipped that a person's life is one long transient, ending at the singularity of death.)

Figure 2.4. A singularity encountered when a filament breaks (ZZM). The "filament" could easily be a whip, and explains why one hears a "snap" when it is cracked—the energy from the wave of "cracking" accumulates and is "terminated" at the end. A similar mechanism explains the sound heard when a string under tension suddenly breaks.

As has been stressed, a fundamental problem in the application of nonlinear dynamics to physiological systems is that fact that they are still classical: they require Lipschitz conditions to guarantee the uniqueness of solutions subject to prescribed initial conditions. For a dynamical system all the derivatives are bounded. This condition allows for the description of classical Newtonian dynamics within the traditional theory of differential equations, such that reversibility and predictability are guaranteed (recall that chaotic dynamics can be predicted if it were possible to know the initial conditions with infinite precision and devoid of noise). This results in an infinite time of approaching an attractor. Clearly, however, such observations are not typical in real systems—especially biological. One way to overcome these problems is to violate the

Singularities and Instability

requirement of Lipschitz conditions. By doing so, at singularities of the phase space, the dynamics forgets its past as soon as it approaches these singularities. Additionally, the dynamics of such systems is able to be activated not only by external inputs and infinitesimal noise of any kind, but also by internal perturbations. Such critical points can be placed in a network with others which are weakly coupled, and can perform parallel tasks.

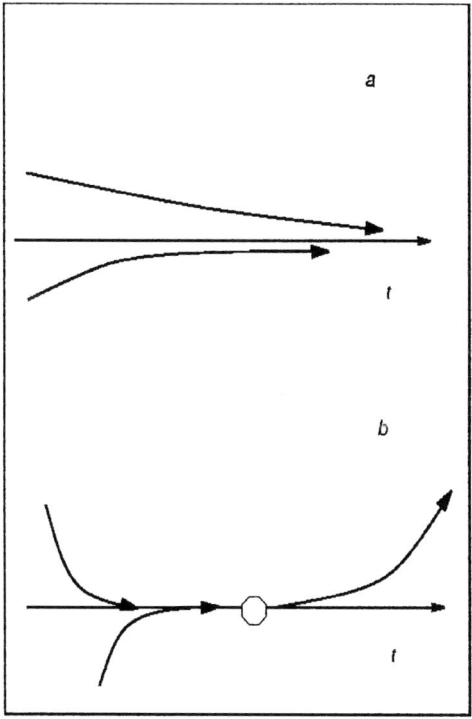

Figure 2.5. Classical dynamics (a) approach an equilibrium in an infinite time; while nondeterministic dynamics ("terminal") dynamics approach an equilibrium singularity in finite time (b), and leave in finite time.

2.4 Lipschitz Conditions

To be more explicit, consider, for example, a simple equation without uniqueness: $dx/dt = x^{1/3} \cos \omega t$. At the singular solution, $x = 0$ (which is unstable, for instance at $t = 0$), a small noise drives the motion to the regular solutions, $x = \pm (2/3\omega \sin \omega t)^{3/2}$ with equal probabilities. Indeed, any prescribed distribution can be implemented by using non-Lipschitz dynamics. It is important to emphasize, however, the fundamental difference between the probabilistic properties of these non-Lipschitz dynamics (N.B., I also call these dynamics "terminal dynamics" and "nondeterministic" dynamics to emphasize different aspects.) and those of traditional stochastic or differential equations: the randomness of stochastic differential equations is caused by random initial conditions, random force or random coefficients; in chaotic equations small (but finite) random changes of initial conditions are amplified by a mechanism of instability. But in both cases the differential operator itself remains deterministic. Thus, there develop a set of "alternating," "deterministic" trajectories at stochastic singularities.

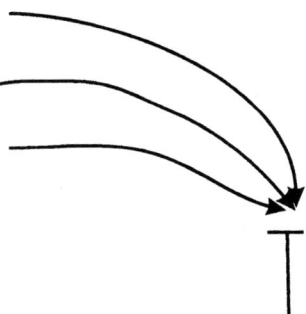

Figure 2.6. Demonstration of nondeterministic dynamics of a hammer hitting a nail. Multiple trajectories arrive at the same point (the nail head), while the arm swinging the hammer also leave the head with multiple trajectories.

One additional note: although there are many models which add stochastic terms, these terms are added to the entire dynamic. Thus the noise could theoretically distort the entire dynamic. The model here proposed would suggest that the stochastic aspects appear only at the singular points, and thus do not distort the general dynamic of the trajectory. Numerous other models (theoretical and real can also be demonstrated (Figs. 2.5, 2.6). The difficulty, of course, is to determine the proper form of the deterministic "trajectories" and their related probability distribution functions. At each singular point, a variety of connecting possibilities are admitted—each with their own likelihood of being realized. Furthermore, the strength and type of excitation conditions these possibilities.

Thus, the final problem is cast in the form of a combinatorial problem, resulting in a "stochastic attractor" (Fig. 2.5).

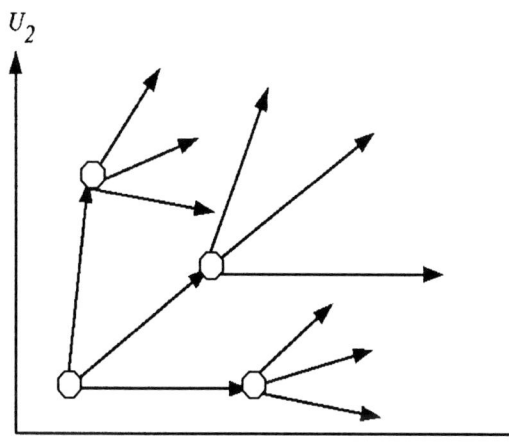

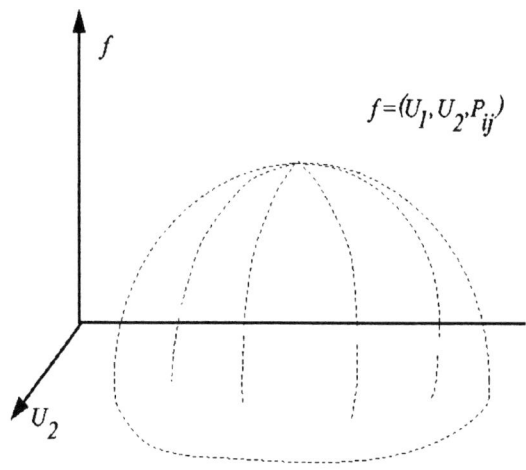

Figure 2.7. Connected singular, terminal dynamics can demonstrate multi-choice probability distributions sensitive to type and kind of perturbation (noise) at each singular point (top). At each singularity, multiple choices are available. Thus one can speak of a "stochastic attractor" generated as the multidimensional probability surface (bottom). Note that this, in effect, reduces the dynamical "problem" to a

statistical "problem," based on combinatorics. The implications for brain dynamics (see below) are considerable.

Table 2.1.

Chaotic Dynamics	Non-Lipschitz Dynamics
Dependent on initial conditions	Dependent on singularity
Attractor present	No attractor in the usual sense
Spreads uncertainty over entire attractor; i.e., global instability	Uncertainty at singular points (stochastic attractor)
Behavior varies	Dynamics is well behaved away from singular points
Smooth stretching and folding of attractor	Random spread of points in region of phase space (stochastic attractor)
Information decreases with time	Information infinite at singularity
Cantor set	Countable combinatorial set
Short term predictability	Predictability based on probability e
Controlled by external forces	Easily controlled at singularity
Time is a continuum	Near singularity, time is decoupled

In contradistinction to this, in non-Lipschitz dynamics, randomness results from the violation of the uniqueness of the solution at equilibrium points, and therefore, the differential operator itself generates random solutions. Furthermore, the singular point can be part of a larger chain of oscillators which become self-organizing. Interestingly enough, when such dynamics are analyzed numerically for Liapunov exponents, positive values are obtained, which is not surprising since the solutions are characterized by an infinite divergence. (Table 2.1). Another important point is the fact that the probabilities are part of a multiplicative process; i.e., the process along an "arm" is a result of the combinatorics of the arm—although each point is independent. Or stated otherwise, for a given singular probability P_0, at each step j, the probability can change according to a variable F_j, such that $P_j = F_j P_{j-1}$. The result of such a process is often described as a lognormal or perhaps a power law distribution. Of some interest is the fact that such distributions are often found in nature—sometimes also possibly fractal. To be sure, other kinds of processes can generate similar processes, which points out the importance of going beyond the description of a distribution, to the generating process itself in order to understand the relevant governing variables.

The important consideration here is that the particular form for these singularities is not unique. There are numerous models which create similarly behaving singularities. Thus, it is crucial to detect such singularities so that apparent complexity is not mistaken for actual complexity in the underlying system. An important feature of such systems is the abrupt change of the trajectory as it changes from one solution to another as it approaches the singularity. This jump is essentially instantaneous. Typically, the divergence occurs at the second (or higher) time derivative. The ramifications for such dynamics may be especially important for modeling biological processes such as the nervous system, where "natural singularities" in the form of synapses exist. A problem, however, remains in detecting such dynamics in real life.

2.5 Basic Concepts

The equations of classical dynamics based upon Newton's laws: are the generalized coordinates and velocities, include a dissipation function which is associated with the friction forces: the functions do not follow from the Newton's laws, and, strictly speaking, additional assumptions need to be made in order to define it. The "natural" assumption (which has been never challenged) is that these functions can be expanded in Taylor series with respect to an equilibrium state. Obviously this requires the existence of the derivatives: i.e., the function must satisfy the Lipschitz condition. This condition allows one to describe the Newtonian dynamics within the mathematical framework of classical theory of differential equations. However, there is a certain price paid for such a mathematical "convenience": the Newtonian dynamics with dissipative forces remains fully reversible in the sense that the time-backward motion can be obtained from the governing equations by time inversion.

In order to trivialize the mathematical part of the argumentation, consider a one-dimensional motion of a particle decelerated by a friction force: the equilibrium cannot be approached in finite time. The usual explanation for such an effect is that, to the accuracy of limited scale of observation, the particle "actually" approaches the equilibrium in finite time. In other words, eventually the trajectories and the equilibrium point become so close that one cannot distinguish one from the other. The same type of explanation is used for the emergence of chaos: if two trajectories originally are "very close," and then they diverge exponentially, the same initial conditions can be applied to either of them, and therefore, the motion cannot be traced.

2.5.1 Dissipation

This paradox can be obviated by a specific formulation of the dissipation which makes the Newtonian dynamics irreversible. The main properties of the new structure are based upon a violation of the Lipschitz condition. This is to say that when the condition is violated, the friction force grows sharply at the equilibrium point, and then it gradually approaches the straight line. This effect can be interpreted as an instantaneous jump from static to kinetic friction. It appears that this small difference between the friction forces leads to fundamental changes in Newtonian dynamics:

- First, the time of approaching the equilibrium becomes finite.

- Second, the motion, has a singular solution and a regular solution.

In a finite time the motion can reach the equilibrium and switch to the singular solution, and this switch is irreversible. It is interesting to note that the time-backward motion is imaginary, whereas the classical version of this motion is fully reversible.

The equilibrium point represents a "terminal" attractor which is "infinitely" stable and is intersected by all the attracted transients. Therefore, the uniqueness of the solution at the equilibrium is violated, and the motion is totally "forgotten." This is a mathematical implication of irreversibility of the dynamics.

So far we were concerned with stabilizing effects of dissipative forces. However, as well-known from dynamics of non-conservative systems, these forces can destabilize the motion when they feed the external energy into the system. In order to capture the fundamental properties of these effects in the case of a "terminal" dissipative force, assume that now the friction force feeds energy into the system: one can verify that the equilibrium point becomes a terminal repeller, and is "infinitely" unstable. If the initial condition is infinitely close to this repeller, the transient solution will escape it during a finite time period, while for a regular repeller, the time would be infinite.

As in the case of a terminal attractor, here the motion is also irreversible: the solutions are always separated by the singular solution, and each of them cannot be obtained from another by time reversal: the trajectory of attraction and repulsion never coincide.

But in addition to this, terminal repellers possess even more surprising characteristics: the solution becomes totally unpredictable. Indeed, two different motions described are possible for almost the same initial conditions. The essential property of this result is that the divergence of these two solutions is characterized by an unbounded rate. Thus, a terminal repeller represents a short, but powerful pulse of unpredictability which is pumped into the system via terminal dissipative forces. Obviously failure of the uniqueness of the solution here results from the violation of the Lipschitz condition.

As is known from classical dynamics, the combination of stabilizing and destabilizing effects can lead to a new phenomenon: chaos (N.B. This is chaos in its most generic sense—not the "deterministic" type.). Here stabilization and destabilization effects alternate. With the initial condition the exact solution consists of a regular solution, and a singular solution. During the first period there is a terminal repeller. Therefore, within this interval, the motion can follow one of two possible trajectories (each with the probability 1/2) which diverge with an unbounded Liapunov exponent. During the next period the equilibrium point becomes a terminal attractor; the solution approaches it and it remains motionless. After this, the terminal attractor converts into terminal repeller, and the solution escapes again, etc.

It is important to note that each time the system escapes the terminal repeller, the solution splits into two symmetric branches, so that there are several possible scenarios of the oscillations with respect to the equilibrium. Hence, the motion resembles chaotic oscillations known from classical dynamics. It combines random characteristics with the attraction to a center. However, in the classical case the chaos is caused by a sensitivity to the initial conditions, while the uniqueness of the solution for fixed initial conditions is guaranteed. In contrast to this, the chaos in these oscillations is caused by the failure of the uniqueness of the solution at the equilibrium points, and it has a well as an organized probabilistic structure. Since the time of approaching the equilibrium point by the solution is finite, this type of chaos can be called "terminal," or nondeterminisitic.

Within the framework of terminal dynamics, formations of new patterns of motion can be understood as chains of terminal attractions and repulsions.

During each terminal repulsion the solution splits into two symmetric branches, and the motion can follow each of them with equal probability. Such a scenario can be represented by terminal chaos, which has an exact mathematical formulation, and does not depend upon the accuracy to which the initial conditions are known. Driven by non-uniqueness of solutions at terminal repellers, terminal chaos, and consequently, the process of emergence of new patterns of dynamical motions, possesses a well organized probabilistic structures.

In conclusion it should be stressed again that all the new effects of terminal dynamics emerge within vanishingly small neighborhoods of equilibrium states which are the only domains where the governing equations are different from classical.

2.5.2 Terminal Dynamics Limit Sets

The equations of classical dynamics themselves may be derived from a Lagrangian function, from variational principles, or directly from Newton's laws of motion, and they may be presented in various equivalent forms. However, one mathematical restriction to all of these forms is always applied: the differential equations describing a dynamical system must satisfy the Lipschitz condition which expresses that all the derivatives must be bounded. Actually this mathematical restriction guarantees the uniqueness of a solution subject to fixed initial conditions. Such a uniqueness seemed to be very important while the dynamical systems have been applied for modeling of energy transformations in mechanics, physics, and chemistry. However, attempts to exploit classical dynamics for information processing with applications to modeling biological and social behaviors have exposed certain limitations of the approach because of determinism and reversibility of solutions. Mathematical and physical aspects of these limitations as well as the consequences of their removal were discussed in the previous section. Here we present a general structure of dynamical systems which does not possess a unique solution due to violation of the condition at equilibrium points.

In order to emphasize the difference between classical and terminal equilibrium points, consider the simplest terminal system. The relaxation time for a solution with the initial condition to this attractor is finite. Consequently, this attractor becomes terminal. It represents a singular solution which is intersected by all the attracted transients the equilibrium point. If the initial

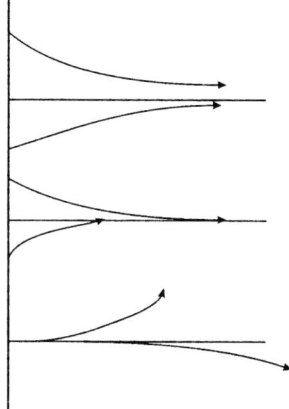

Figure 2.8. Terminal attractors and repellers exhibit different trajectories in time. A traditional attractor never really meets the equilibrium—even to infinity. Whereas a terminal attractor (middle) and repeller (bottom) devolve end evolve in finite time. The actual time series of terminal centers appear at different points in the series depending upon the length of time present at the singularity. For this reason, a naïve inspection of the series may provide a false impression of the dynamics structure.

condition is infinitely close to this repeller, the transient solution will escape the repeller during a finite time period, while for a regular repeller, the time would be infinite.

2.5.3 Interpretation of Terminal Attractors

As has been pointed out in the introduction of the chapter, the mathematical formalism of terminal dynamics follows from a more general structure of the dissipation function, which allows for the existence of smooth transitions from static to kinetic friction. It should be emphasized that the behavior of the solutions around the equilibrium points in terminal dynamics is more "realistic" than in classical dynamics since actual time of convergence to equilibrium points is finite (Fig. 2.8). However, in order to make it finite, one has to violate the Lipschitz condition since all the trajectories must intersect at the equilibrium point. In classical dynamics, the Lipschitz condition is not violated, and the infinite time of convergence is explained in terms of "small dissipative forces" which are always present. Actually terminal dynamics incorporates these forces which can be found from measurement of the convergence.

It can be shown that the terminal attractor as a mathematical concept has other physical interpretations, and one of them is the energy cumulation effect, although in this case one deals with the finite time of convergence of a propagating wave rather than a motion of an individual particle. As an example, consider a propagation of an isolated pulse in an elastic continuum. The dynamics of the pulse propagation has a terminal attractor. But if the leading and the trailing fronts of the propagating pulse approach the same point during finite time, then eventually the width of the pulse will shrink to zero, and all the energy transported by the pulse will be distributed over a vanishingly small length. Hence, the existence of a terminal attractor in such models leads to an unbounded concentration of energy in the neighborhood of the attractor.

Based upon this model, one can explain and describe the formation of a snap at a free end of a suspended filament, as well as the cumulation of the shear strain energy at the soil surface as a response to an underground explosion (earthquake). In these models, the free end of the filament and the free surface of the soil serve as random attractors.

2.5.4 Unpredictability in Terminal Dynamics

The concept of unpredictability in classical dynamics was introduced in connection with the understanding of chaotic motions in nonlinear systems.. Such motions are caused by the Liapunov instability, which is characterized by a violation of a continuous dependence of solutions on the initial conditions during an unbounded time interval. That is why the unpredictability in these systems develops gradually.

Consider the transient escape from a terminal repeller: two different solutions are possible for "almost the same" initial conditions. The most essential property of this result is that the divergence of the solutions is characterized by an unbounded parameter: here the rate of divergence can be defined in an arbitrarily small time interval, since during this interval the initial infinitesimal distance between the solutions becomes finite. The resultant dynamics defy predictability

2.5.5 Irreversibility of Terminal Dynamics

Classical dynamics describe processes in which time plays the role of a parameter: it remains fully reversible in the sense that the time-backward motion can be obtained from the governing equation by time inversion. This means that classical dynamics cannot explain the emergence of new dynamical patterns in nature.

However, there exists another class of phenomena where past and future play different roles, and time is not invertible: by definition (the second law of thermodynamics) irreversibility is introduced in thermodynamics by postulating the increase of entropy. As stressed by Prigogine, irreversible processes play a constructive role in physical world; they are at the basis of important coherent processes that appear with particular clarity on the biological level. In terminal dynamics one to one mapping between past and future may not exist, and this is why the definition of irreversibility should include (both conservative and non-conservative) systems whose past can be uniquely reconstructed from the future in contradistinction to those systems for which this cannot be done.

2.5.6 Probabilistic Structure

As shown above, the terminal version of Newtonian dynamics is different from its classical version only within vanishingly small neighborhoods of equilibrium states, and therefore, it contains classical mechanics as a special case. This means that terminal dynamics is not necessarily always unpredictable and irreversible: in some domains it is identical with the classical dynamics. However, here attention will be focused on specific effects of terminal dynamics, and in particular, on its probabilistic structure.

There exists a fundamental difference between the probabilistic properties of terminal dynamics and those of stochastic or chaotic differential equations. Indeed, the randomness of stochastic differential equations is caused by random initial conditions, random force or random coefficients; in chaotic equations small (but finite) random changes of initial conditions are amplified by a mechanism of instability. But in both cases the differential operator itself remains deterministic. In contradistinction to this, in terminal dynamics, randomness results from the violation of the uniqueness of the solution at equilibrium points, and therefore, the differential operator itself generates random solutions.

It should be emphasized that this is a phenomenon which does not exist in classical version of nonlinear dynamics. Unlike chaotic attractors, here the probability density can be uniquely controlled by the parameters of the original dynamical system, while the limit stochastic process does not depend upon the initial conditions if they are within the basin of attraction. The density can also be controlled by coupling with other similar dynamics. If this happens, the connected system develops a complicated combinatorial system. Not only can activity be enhanced, it can also be extinguished. The obvious similarity with brain neurodynamics.

2.5.7 Self-Organization in Terminal Dynamics

A dynamical system is called self-organizing if it acquires a coherent structure without specific interference from the outside. Terminal dynamics possesses a powerful tool for self-organization coming from a possibility of coupling between the original dynamical system and its own associated probability density dynamics.

The solution to a terminal dynamical system approaches a stochastic attractor with a given probability density. It should be stressed that this attractor has not been "stored" in prescribed coefficients of: the dynamical system "finds" it as a result of coupling with its "own" probability equations. In general, parameters of the dynamical system can be coupled with the moments of the probability density, and that will lead to new self-organizing architectures.

Although the term "self-organization" is used here, I differ somewhat from a common interpretation; namely, that this self-organization develops organically. I believe there is no evidence to suggest that such self-organization must develop given a certain type of dynamics. This is to say that a certain dynamics exhibits a "teleological" activity. Because the dynamics are divorced from the past, self-organization qua determinism is not an option, but rather a possibility. And even then, the "organization" has no requisite persistence. Because everything "works" via probabilities, there are no inherent guarantees that certain "structures" will be seen.. Indeed, the question has to be asked whether the self-organization is perhaps more of a human desire to see patterns of organization

This point is crucial to understanding so-called "self-organizing" systems. The particular view adopted exhibits profound implications for the ultimate understanding as well as so-called control of such systems. Frequently, these "systems" are systems only because they have been defined as such. Their function and boundaries from an epistemological view are more often than not a convenience for understanding and study. It should be recognized, however, that this is a mental convenience, which does not necessarily conform to the reality. Because the mental aspects may be so significant, accrued connotations regarding systems are easily applied to the phenomena. Specifically, these often concern determinism and all that this term implies.

As has been already suggested, deterministic dynamics define the motion of the dynamics for all time. Implicit in this formulation is that there is a predestination circumscribed by the system boundaries. With non-deterministic dynamics this is not a reality. The fact that a "system" develops is totally dependent upon probabilities: probabilities might be just as good not to form a system.

It should be recalled that frequently, evidence for "self-organizing systems" is presented in cases where there is evidence for this. Negative evidence (falsification) is hardly ever presented, because, of course, it is not interesting. This is certainly a situation which begs for classical "controls." The fear of the absence of positive findings may be a motivating factor, yet if these "systems" are approached from the possibility of nondeterministic dynamics, the situation may be far from negative. It may be that these systems can exhibit the so-called self-organization only probabilistically, provided appropriate boundary conditions exist. And certainly, experimental boundary conditions may give a false impression when the real contextual reality is considered.

There are increasing attempts to view metabolic processes as part of networks. Yet, many of these networks often require constant modifications because of the sheer numbers of components involved.

A more subtle question involves the boundaries of these networks. Consider, for example, the difficulty of establishing atmospheric models: gradually, it has become understood that the atmosphere is constrained by ocean dynamics, which are also modified by geological dynamics. This is not even to consider cosmological dynamics such as solar winds.

This is not to say that such conceptions are improper; rather, the intent here is to suggest that nondeterministic perspectives may provide a less constraining framework.

3 *Noise and Determinism*

When, almost two decades ago, "chaos theory" emerged as a possible solution for explaining biological phenomena, many scientists believed that the theory could give an efficient description of complex biological systems by the use of so called "invariants," i.e., measures computable by the analysis of experimental observables that would give a description of a some sort of "essence" of the studied system, independent of what was considered a mere consequence of contingencies like the adopted measurement scale, detection instruments and so forth. This concept of "invariant" is a basic tenet of the way physicists look at reality, and implies the possibility of making a neat separation between what is essential (and thus repeatable, rule-obeying and thus interesting) and what is contingent, occasional, noisy and simply disturbing. Chaos theory was seen as a positive development by many scientists because it gave the promise of putting order in the wild territories of contingencies and noise.

The last decade, on the contrary, has witnessed just the opposite in the analysis of biological signals: the original hope that "chaos theory" would help elucidate the complexities of biology are being questioned. As more investigators became aware of the mathematical requisites (and limitations) of chaotic measures such as Liapunov exponents and dimensions, they recognized that new tools are needed. An important recognition in this respect is that biological signals, in addition to being nonlinear, tend to be nonstationary, noisy and high dimensional. Certainly such a statement is not revolutionary, however, during a time when new, exciting concepts are emerging, it sometimes becomes easy to overlook basic facts, and to ignore fundamental assumptions.

Our own doubts about the ability of many chaotic measures to clarify physiologic processes surfaced with work to understand heart rate dynamics. Even cursory examinations of plots of heart rate demonstrate frequent, often sudden transitions. (In this sense, the experience was similar to the experience with Fourier transforms: they are linear tools which presume data stationarity.) A now classical approach to such a problem was as follows: to reduce the putative degrees of freedom, correlation dimensions were calculated for heart transplant recipients, whose hearts, by virtue of the surgery, were denervated. Theoretically, denervation reduced the complexity of factors governing the control of the heart. The subjects were equilibrated to a quiet environment and resting for several minutes prior to recording. Surprisingly, stationarity was exceedingly difficult to obtain, and, moreover the calculated dimensions were inconsistent and relatively high. In an attempt to gain greater control, experiments were performed on isolated, perfused rat hearts; i.e., hearts not connected to an organism. Again, difficulties were observed in gaining the requisite stationarity, and dimensions and entropies exhibited error bars which could not confirm chaotic dynamics. More importantly, a piecewise linear map was successful at modeling the dynamics, but only with the addition of a small amount of noise to force the dynamics (see below). An important feature of this map was the necessity of introducing noise to force new trajectories—otherwise the dynamics would remain in a quiescent state. Although the interpretation at that time was that of "noise induced intermittency," clearly an alternate interpretation is that of transient trajectories, which return to a "stable node," only to be forced out again by the noise.

The requirement that noise be an important force for the dynamics was not limited to cardiac dynamics, or to simple one-dimensional maps. We obtained similar results in the modeling of cat phrenic nerve activity, but this time using modified Bonhoeffer-van der Pol (BVP) equations: for a certain set of parameters, trajectories converged on a "node"; by the addition of noise, the dynamics were forced off the node and back to the node by a circuitous route (Fig. 3.1).

Careful consideration of the BVP equation indicates why this is so. Specifically, the equations

$$\frac{dx}{dt} = c(y + x - x^3/3 + z) \quad (3.1)$$

$$\frac{dy}{dt} = -(x - a + by)/c \quad (3.2)$$

were put forth by FitzHugh as generalized van der Pol equations, as a simplification of the Hodgkin-Huxley equations for excitable-oscillatory systems. He suggested that x represented membrane potential; y the original n and h; and z a current stimulus.

Investigators have determined that as is trajectories go upward, trajectories merge into a separatrix. Stable equilibrium is found at the intersection of x and y nullclines. Depending upon the parameters, the attractors are either a stable node or limit cycle. In fact more research has indicated that the nodes may be unstable depending upon a very narrow tuning of these parameters. Thus, depending upon the refinement of the calculation (level of noise)—the result can be totally different (See below relative to the neutron star equations the discussion of "practical" appreciation of equations.) Interestingly, in dealing with phase resetting experiments with cats by the perturbation of mesencephalic stimuli, it was noted that unpredictable intermittent oscillatory activity resulted. The BVP model demonstrated this kind of activity through adjustment of a noise parameter.

Interestingly enough, we found that with increasing noise amplitude, the frequency of the oscillations became more regular, i.e., an example of stochastic resonance. As with the cardiac data, we found that although each of the oscillations was a transient, single trajectory, it was possible to obtain a fractional dimension as well as a positive Liapunov exponent (4.68 and 0.162 respectively). Thus, an immediate discrepancy became obvious: using typical techniques, "fractal" dimensions could be obtained, yet, clearly, the dynamics of the process did not meet the requisites of the presumed system. Thus, although one could obtain these odd measures, they, in themselves, were not sufficient for justification of the system as being "chaotic."

These results were at one time both interesting yet dismaying: clearly, by definition, dimensions and Liapunov exponents are the properties of continuous systems, yet these "transient" dynamics were capable of generating so-called chaotic measures. Some of these "chaotic" results can be relegated, in part, to the well known problems associated with the calculation of dimensions and exponents of experimental data. Yet a more fundamental question centers on the correct characterization of the dynamics: are they smooth and continuous, or are they essentially discontinuous transients? And why does noise seem to be so important? Careful consideration of the involved physiology, would suggest that "transient" dynamics are the correct model for real organisms, in that real organisms must constantly change and respond to environmental factors, whereas continuous dynamical models would limit this ability. These questions led us to reconsider Newton's formulation of deterministic dynamics.

Unstable Singularities

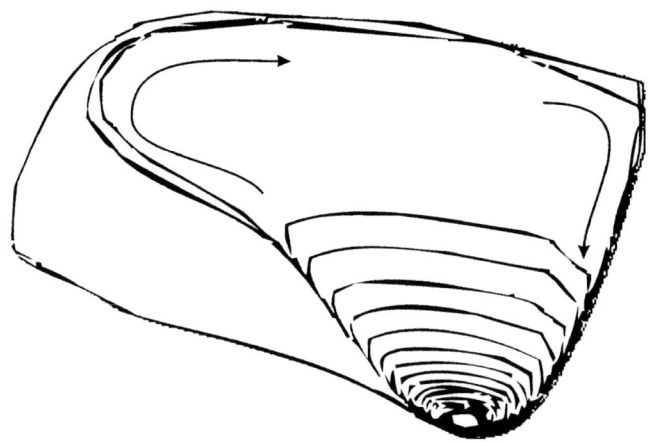

Figure 3.1. Composite phase plane diagram of BVP Model for several different parameter values. Note the "resting point," or "stable node." In reality, very small amounts of noise prevent any "rest." Instead, the noise forces the dynamics off the node and along on of the many possible trajectories which return to the node, only to be forced off again. Thus the noise is responsible for the dynamics. The strength of the noise determines the probabilities for the various trajectories. Although the BVP model is highlighted, there are many such models, and many "one-hump" maps which describe similar dynamics. They have been especially popular to describe various physiological properties; however, they should be regarding as purely phenomenological since the correspondence with actual physical variables remains elusive.

According to Newton's principle of determinacy, all motions of a system are uniquely determined by their initial positions, and initial velocities. Newtonian determinism counterdefines nondeterministic dynamics as systems whose initial conditions do not uniquely determine the state of a system at all later points in time. Such dynamics, however, should not to be confused with chaotic dynamics: chaotic dynamics are deterministic in that their motions can be predicted (in principle) if their initial conditions were known with infinite precision. Additionally, deterministic chaotic dynamics require trajectories which exponentially diverge as given by positive Liapunov exponents. Nondeterministic dynamics, on the other hand, cannot be predicted for the future even if infinite precision were possible, and no noise present, since such dynamics consist of a single trajectory. The explanation for such dynamics is that solutions to the dynamics are not unique for all time, and indeed have multiple solutions (branching) or merging trajectories (Figs. 3.2, 3.3). This is to recognize that Euler's equations of motion cannot be made explicit so as to satisfy a Lipschitz condition, (i.e., a unique solution) and instead, uniqueness must be studied case by case. Many will recall the fact there are really very few differential equations describing natural phenomena. And in the engineering perspective, many solutions to practical problems require ad hoc methods employing, for example, finite element methods.

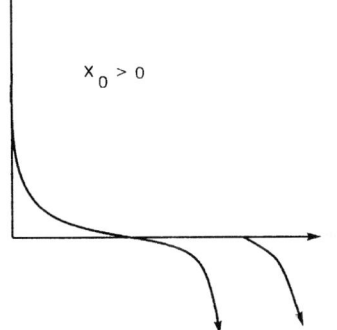

Figure 3.2. Simple example of divergence with multiple trajectories for a particle with equation of motion, $\dot{x} = -c\sqrt{|x|}$. Lipschitz conditions are not met at $x = 0$, and are nondeterministic: $u(t) = c^2/4(t-t_1)^2$ for $t \leq t_1$, 0 for $t_1 \leq t \leq t$, and $-c^2/4(t-t_2)^2$ for $t \geq t_2$, where t_1 and t_2 are points where solutions meet and leave the time axis, and where possibly $t_1 = -\infty$ and/or $t_2 = +\infty$. And if $x_0 \neq 0$ then for initial conditions $t_0, x_0, t_1 = t_0 + (2/c)\sqrt{x_0}$ for $x_0 > 0$, and $t_2 = t_0 + (2/c)\sqrt{-x_0}$ for $x_0 < 0$.

The motion of macroscopic quantities of a macroscopic system is deterministic if both the approximate inertial manifolds and the flow vector field on the approximate inertial manifolds are smooth (i.e., continuously differentiable) which is generally not the case in physical and biological systems. Indeed there are strong indications that for most experimental systems the motion of the asymptotic inertial manifold is not smooth, but discontinuous and nondeterministic. Even if the dynamics of the full system is deterministic, and even if the neglect of some rapidly decaying variables is well justified through time scale arguments, there is no reason to assume that the property of determinism is preserved, if one or several variables in the equation of motion are neglected. An example of this argument familiar to everyone is the motion of a flag.

It can be assumed that the front part of a flag is at rest whereas the rear end is bent around and moving with velocity v_0. If we ignore the vibrational states, a relatively simple model representing the conservation of energy would be

$$m\dot{x}^2/2 = \text{const}, \qquad (3.3)$$

where x is the center of mass of the moving part of the flag, $m = (2/3)\rho(l-x)$ is the mass of the moving part, ρ is the mass density, and l the length of the flag. If the equation is rewritten as

$$\dot{x} = \sqrt{3\,\text{const}/\rho(l-x)}. \qquad (3.4)$$

The solution of this differential equation end at $x = l$. In real systems the results are informative: all kinetic energy is stored at the free end of the flag and therefore the forces become so big that the end of the flag breaks away. This means that the trajectory of the asymptotic inertial manifold is terminated. Vibrational modes are slaved before they are stimulated and can produce sound and damage to the flag.

A unique consequence of nondeterministic dynamics is their behavior in the presence of noise. This can be seen in the context of an experiment involving a sphere in castor oil slowly sliding down a specially milled slope (Fig. 3.4). The potential describing this is

$$V(x) = -(c/n)|x|^{n-1}, \qquad (3.5)$$

with $1 < n < 2$, and

$$\dot{x} = c|x|^{n-1} + F(t), \qquad (3.6)$$

where $F(t)$ is uncorrelated white noise with standard deviation of D. In the no noise case, and $1 < n < 2$, the trajectories intersect at $x = 0$ (nondeterministic dynamics), but for $n > 2$, the solutions are unique. Also at $x = 0$ for all $n > 1$ there is a saddle point (a point whose trajectories describe a saddle shape). In the nondeterministic case, the system reaches the saddle point in a finite period of time, but in the deterministic case the mechanics get closer and closer to the point without ever reaching it. The addition of noise to the system further separates the dynamics: the deterministic system passes the singularity when the noise amplitude is approximately equal to the distance from the point. For nondeterministic systems, however, the orbit reaches the singularity in a finite time (Fig. 3.5) Then, a minimal noise kicks it out. I emphasize the existence of the singularity which admits multiple branching solutions, and at the same time the strong influence of minimal amounts of noise. (I also note that this is an important distinction between topological singularities such as so-called "unstable periodic orbits" and the singularities discussed here. Topological singularities are a result of the mathematical formalisms involved in defined vector fields on an inertial manifold. As pointed out above, this is not the case with biological motions.)

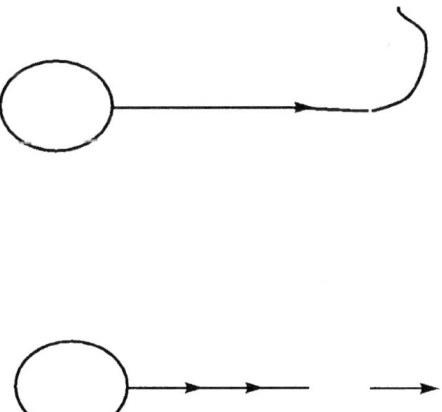

Figure 3.3. Diagram of a flag tethered at the circle (top). The vibrational modes are abruptly terminated at the flag boundary resulting in a singularity and a blow up physically manifested by a tear or rip.

A biological example deals with an experiment with arm motion, where it has been shown that oscillators of the van der Pol (VDP) type:

$$\ddot{x} + \eta(x^2 - 1)\dot{x} + p_1 x + p_2 x^2 + p_3 x^3 + \ldots = F(t) \qquad 3.7$$

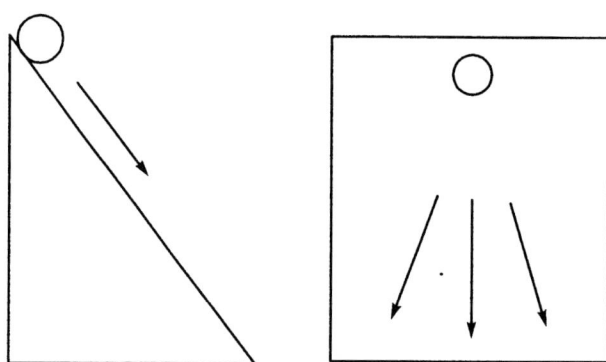

Figure 3.4. Cartoon of sphere on slope. Right, depending on force and exponent, the ball rolls down a single, or can roll down multiple different paths.

are well suited to describe the global symmetry of the limit cycle and the response of the motions to external perturbations, However, for $F = 0$ the dynamics of the VDP oscillators are always periodic, whereas the natural motions of the arm are not. Instead there is a slight pause of the motion at extreme locations and after a rather undetermined period of time starts to move again. This means the VDP oscillator does not adequately model the interference of putatively higher neural inputs. A careful analysis of the trajectories shows that they have a kink at the most extreme locations. A model which uses one VDP for motion in one direction and another for motion in the opposite direction can perfectly model the global geometry as well as the kink near the ends. However, the corresponding motion is no longer deterministic at $\dot{x} = 0$. After a finite period of time, the smallest control force (e.g., noise) will make the arm move. This type of motion can be approximated by many typical motions such as that of a forearm and hand directing a computer mouse—or an arm waving. (Section 4.4 below demonstrates this explicitly with some experimentally recorded data.)

3.1 *Experimental Determinations*

To determine if a given experimental system follows deterministic versus nondeterministic dynamics we first convert the oscillations to action angle coordinates by normalizing the oscillations to an amplitude of one. This allows

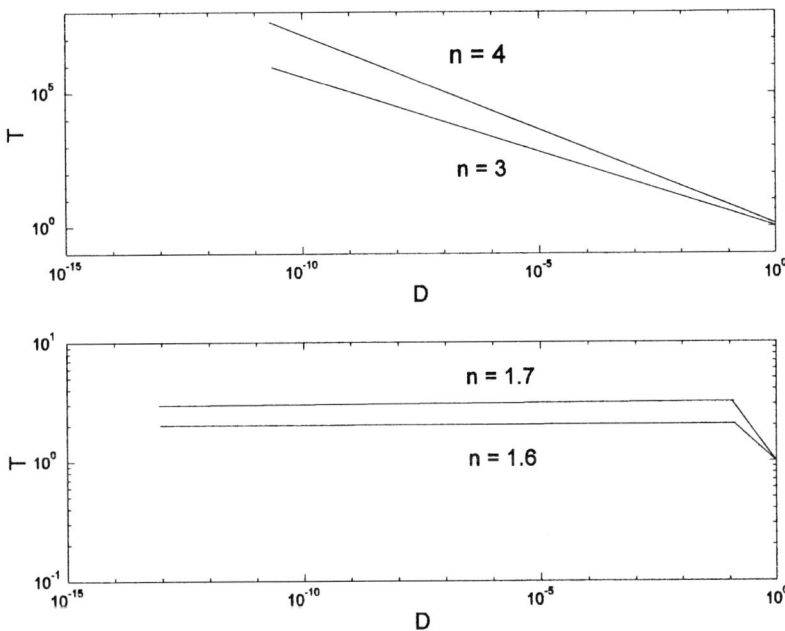

Figure 3.5. Numerical study of first passage time (T) vs. noise (D). The smaller the noise, the longer it takes to pass the saddle (top). For the nondeterministic case (bottom) the system reaches the saddle in a finite time.

us to discard the amplitude information while retaining phase information (this in effect reduces the system by one degree of freedom). Next the phase space is reconstructed using the method of time delays. By inspecting the turning point in this space one can determine if the solutions to the system are nondeterministic: nondeterministic solutions are characterized by orthogonal embedding vectors parallel to the coordinate axis system at this point; deterministic solutions, on the other hand do not demonstrate this orthogonality (Fig. 3.6). To further confirm this possibility, in the case where the reconstruction is similar to an oscillator, the data can be fitted to a nondeterministic differential equation:

$$\dot{\phi} = (1 - \phi^2)^\alpha. \tag{3.8}$$

Unstable Singularities

For $=\alpha=0.5$, the equation has three solutions:

$$\phi = 1, \phi = -1, \phi = \sin(t-c), \quad (3.9)$$

where c is a constant. For $\alpha < 1$, ϕ reaches ± 1 in a finite time, and is characteristic of nondeterminism. Once the dynamics have reached the singularity, additional rules are necessary for the dynamics; i.e., stay in the singularity or continue on with motion. In the case of a deterministic oscillator the rule is Newton's second law:

$$\ddot{\phi} + \phi = 0. \quad (3.10)$$

Thus physiological systems which behave as a simple oscillator may be tested for nondeterminism: $\alpha < 1$, if the system is nondeterministic and conversely for $\alpha > 1$.

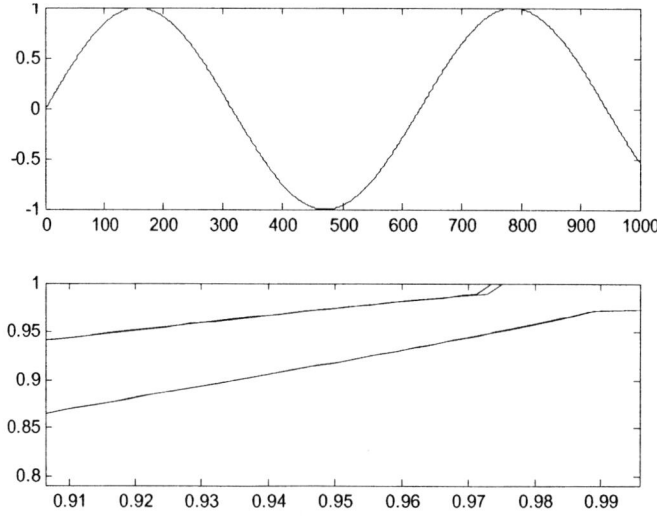

Figure 3.6. Sine wave (top) with singularities at positive value of 1. Delayed vectors (bottom) of sine wave. Note the orthogonal vectors at extreme right indicating independence.

3.2 The Larger Metaphor

It is admitted that the metaphor of "system," i.e., of something that, at least in principle, could be defined by a finite description (a set of rules, a set of differential equations or a probability distribution) that holds true for the period of observation and operates as a sort of "generating function" for its behaviour is tenable for all the objects we study. This will allow us to compare the proposed methodological and theoretical approach with other competing views like deterministic chaos. But it is important to stress this hypothesis is by no means to be considered as generally true. On the contrary, for example, the great majority of biological objects cannot be considered as systems: protein sequences have no generating function other than historical accidents that gave rise to the observed structures over millions of years; and organs are constantly changing their way of operation under the agency of age, pathologies and again historical trajectories. Probably the emphasis on systems derives from the fact that science is familiar with machines which have operating instructions and finite descriptions. Biological entities, on the contrary, are more similar to old cities that can obviously be characterized by maps, but are full of elements doing different jobs, secrete passages and, more importantly, grow and change continuously (albeit often in an interrelated way).

A dynamical system is a system that changes over time. For a dynamical system determinism is the proposition that each event is necessary and a unique consequence of past (future) events. All the events evolve in a temporally ordered sequence and fixes each successive (past) event on the basis of the preceding (future) events in accord with a definite rule.

Determinism became the incorporated basic principle of science starting with 1500 A.D. establishing as a universal law of nature that "cause-and-effect" completely governs all motions and structures at every level of physical reality. Starting with the last 300 years our physical and epistemological knowledge has been dominated by this deterministic model of science. Newton discovered a concise set of principles expressible in only few sentences which he showed could predict motion in an astonishingly wide variety of systems to a very high degree of accuracy. Today determinism still remains the core philosophy of the principles of physical science.

Around 1500 A.D. science also introduced the important innovation that the laws of physical reality could be understood meaningfully, expressing physical properties as quantified measurements. This is to say that they must be expressed by using numerical quantities. This is the reason why the laws of physics are usually expressed by mathematical equations.

The theory of deterministic dynamical systems may be considered a branch of the theory of differential equations. A model is essentially a mathematical construct realized often by differential equations to "explain" a deterministic dynamical system. A deterministic model commonly implies a relationship between current and past states of the system so that predictions about the future course of events for the system may be made with accuracy. Certainly every student has been taught that one of the hallmarks of science is to be able to make predictions of natural phenomena.

The basic tenet is the acceptance that all the macroscopic phenomena follow the classical laws of physics as Newton's laws and Maxwell's equations which, in fact, are differential equations defining dynamical systems. In this context the concept of state of a dynamical system becomes important in order to predict its future behavior. In this manner, the physical and biological phenomena are consequently represented by mappings between input signals, states, and output signals. Nonlinear differential equations are often the most useful method for modeling a given system.

In this view, of the deterministic approach rules are sovereign. It delineates a picture of natural reality all based on the acceptance of the general principle that physical reality unfolds in time like the working of a perfect mechanistic device, running rigidly, without exceptions for shared randomness or deviations from predetermined laws. This is in particular an ontological approach of ingenious realism that refuses any suspension of judgment and attributes to outside reality the intrinsic ability of autonomous and mechanistic time evolution. The ontological aspect is here emphasized since it is a philosophical proclivity net necessarily justified by facts. Certainly, deficits in this view started around 1900 by the formulation of a larger set of physical principles and laws. The success of quantum mechanics and its confirmation opened different questions and, in particular, posed the problem of reconciling deterministic behavior at the macroscopic level with intrinsic indeterminism of physical reality at microscopic levels.

In addition, the development of the theory of chaos in the past two decades suggested a new radical approach to determinism and also indicated a possible resolution of discrepancy between mesoscopic global order and aperiodic random activity at microscopic levels. Today it is largely accepted that chaotic deterministic systems are capable of dramatic changes in their aperiodic states with small changes in their bifurcation parameters. Such systems are low dimensional, stationary, autonomous and essentially noise free. Deterministic chaos governs only a small subset of chaotic systems. Reaction-diffusion processes including chemical morphogenesis, and irreversible thermodynamics constitute specific examples of systems giving order from disorder.

The admission of determinism as a universal and definitive tenet of science, was supported from corresponding philosophical and theological speculations. This was a consequence of limited concern regarding the link between physical reality on the one hand and its mathematical counterpart on the other, represented, as previously indicated, by differential equations theory and existence and uniqueness theorems .Certainly that these laws in general "worked" precluded their serious questioning.

The examination of the conditions posed on differential equations suggests that Nature does not exhibit determinism spontaneously as a universal rule. We must impose some mathematical restrictions from the outside on the nature of the function, whose absence assures that none of the above statements may be true. It follows that determinism is subject to particular conditions having suitable and definite mathematical counterparts in the theory of differential equations. For a deterministic approach or, equivalently, for a state space theory of differential equations, this particular physical case corresponds to establishing conditions a differential equation, subjected to given initial conditions, has a solution that is unique and continuously dependent on the fixed initial conditions. This imposition of "conditions" logically qualifies its universality. Thus, in conclusion, determinism does not follow in a necessary and universal form from physical reality but it follows from the mathematical properties that are admitted from the outside as necessary mathematical conditions for physical and biological systems.

As an alternative, a different kind of dynamics can be considered: solutions of equations for the considered model dynamics evolve deterministically through most of phase space, but, in presence of noise, make nondeterministic jumps to other solutions when the trajectory passes near a singularity in the equations of motion. It emerges also that a nondeterministic chaos should be possible in such a class of physical and biological systems.

It has become clear that low dimensional chaos could not explain the phenomena. Certainly others were beginning to come to similar conclusions, and efforts were initiated to explore the use of methods such as wavelets, surrogate testing, and other so-called nonlinear methods. My perception of these methods, however, was that they were inadequate in that they were often still based on linear systems theory, or required stationarity. Even efforts aimed at developing hypothesis testing, such as surrogate analysis, still require stationarity, which, in my experience, are exceedingly rare in biological systems.

It became clear also that what was going on was an attempt to fit the observed data into the mathematical paradigms; i.e., the paradigms were becoming a Procrustean bed. Data were being detrended, filtered, etc. in order to accommodate the mathematical formalisms of the concepts. The results were highly unreliable. Although clear results often ensued, after the processing, there was no clear idea as to what the data had become, or what correspondence there was to the original data/questions.

It was at this point that a hard look was taken at the biological phenomena of interest. This time, however, the look re-started from the data as they emerge from instruments, without superimposing any mathematical formalism which inevitably imposes an entire world of rules, visions of the world, or theories. At that time, heart beat fluctuations were the object of investigation. In describing the phenomena based on the physiological principles governing them, it was noted that:

- they were discrete events—not continuous;

- they were adaptable (able to adjust to beat to beat changes in the milieu;

- they were easily controllable; and

- they functioned under relatively high levels of noise (although they were not always corrupted by it).

Comparing actual data and the models used to explain them, the basic point was considered; namely, the correspondence of biological data to the fundamental assumptions of chaos and other fully determined models. Specifically, it is noted that chaotic dynamics are continuous, deterministic systems. Although the time series of such systems are random looking, processes which generate them are not: their behavior is, by definition, rigid. Biological systems, on the other hand, require stability, yet at the same time adaptability in the face of changing environmental needs. Although these facts are obvious, the consequences are seldom appreciated. To suggest that physiologic data is chaotic, would assume that such systems are ultimately very unstable globally (This emphasis is made since singularities themselves are unstable, but the global system is not.). This comes as a result from the recognition that the way to change a chaotic system is to change its control variable or by creating specific perturbations, but, as is known, very small changes can produce significantly different behavior. As a result much energy has been devoted to analyzing various methods of calculation.

Insofar as noise is ubiquitous in biological systems, it would seem that instability might be a significant problem for such systems. Furthermore, it would seem that considerable energy might be required for an organism to maintain such control variables. Even if the noise is controlled, the performance of the system is fully prescribed by the initial conditions. To change the system would require external input to overcome the inertia of the past. Yet, experimental evidence clearly demonstrates that most organisms can easily adapt within a range of parameters, and exhibit considerable variances within stationary bounds. Also, it is noted that performance in neural control systems is often parallel and/or asynchronous—processes which do not have adequate models in deterministic nonlinear dynamics. Finally, there is a subtle problem dealing with time scales: chaotic systems are by and large insulated from the microscopic world by a large difference in scale, whereas biologic systems are integrated from molecular to macroscopic.

It was clear from the first two points that "chaotic" dynamics could not be an adequate model. Chaotic dynamics are continuous, and determined from initial conditions. And although there have been various attempts at explaining the discrepancies, clearly they were Procrustean. It is highly doubtful that any purely deterministic system can adequately describe biological behavior.

Why such ideas captured the imagination of (especially) biological scientists is not clear. Certainly a frustration in the inability to understand organisms and their dynamics may be partly a cause, but also a failure to understand and be impressed by relatively nascent chaos (and related) theories added to the difficulty: within a decade wholesale adoption of such ideas into biology occurred. To be sure, some of the difficulty lies with mathematicians and physicists who were too eager to explore new territories without fully understanding them. And finally, it may be because of human psychology which tends to be impressed with complicated mathematics—irrespective of their correspondence with reality (as psychologists suggest, the mind can see patterns where there are none). Surely the reasons are manifold and related to the culture of science itself. This is not to say that mathematics and physics should not be used by biological sciences; however, Einstein's dictum should all the more be recalled: "So far as the laws of mathematics refer to reality, they are not certain. And so far as they are certain, they do not refer to reality."

3.3 Non-Equilibrium Singularities

Heretofore, I have been discussing singularities in the context of equations of motion with an equilibrium as a possible paradigm. However, this need not be the case. As masterfully explored by Dave Dixon, equations of motion can also fail due to a failure of uniqueness where these equations are non-zero.

3.3.1 Simple Harmonic Oscillator

First, we can describe a simple harmonic oscillator (SHO), described by the equations

$$\frac{d}{dt}x = y, \qquad (3.11)$$

and

$$\frac{d}{dt}y = -x. \qquad (3.12)$$

Solutions to these equations are circles in the (x,y) phase plane. Clearly, this oscillator is a deterministic system, so that every point belongs to a unique solution described by a circle with a particular radius

$$r = \sqrt{x^2 + y^2}. \qquad (3.13)$$

Suppose, however, that we apply the following nonlinear coordinate transformation to the SHO phase space:

$$x \to x - r = x - \sqrt{x^2 + y^2}. \qquad (3.14)$$

This translates all points on a circle of radius r in the positive x-direction by an amount equal to r. A family of circles concentric about the origin in the original space will now share a common tangent point at the origin of the transformed

space (see Fig. 3.7)

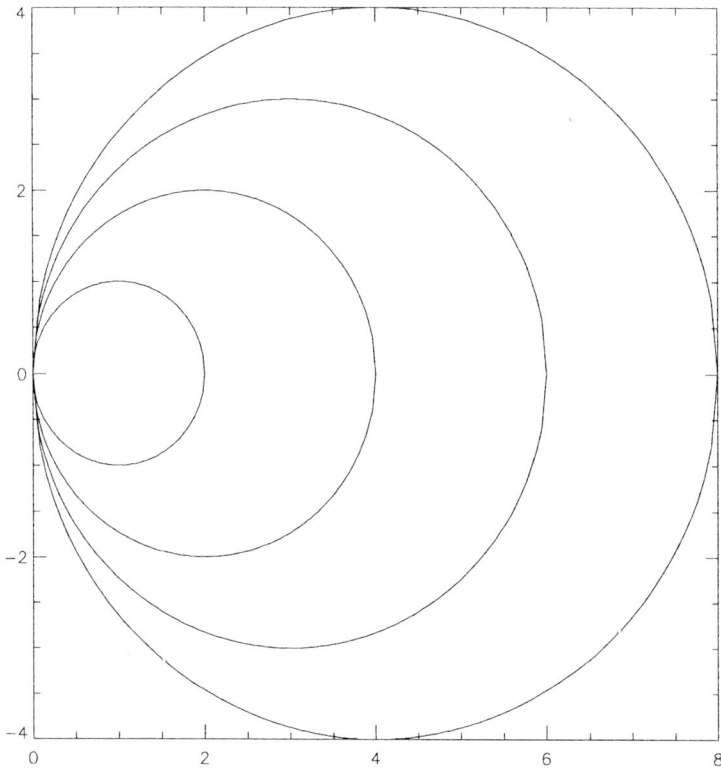

Figure 3.7. Circular orbits sharing a common tangent.

The key feature of this transformation is that some subset of the original phase space decreases in topological dimension as a result of the transformation. For the case here, the entire x-axis is mapped onto a point. And as will be shown, this type of "infinity-to-one" mapping is reflected in the behavior of the transformed dynamics.

The equation of a transformed circle in the new space is given by

$$(x-r)^2 + y^2 = r^2, \qquad (3.15)$$

or, solving for r,

Unstable Singularities

$$r = \frac{1}{2x}(x^2 + y). \tag{3.16}$$

Using this equation, we can apply the transformation to the SHO. The transformed SHO equations of motion in the new coordinate system are given by

$$\frac{d}{dt}x = y, \tag{3.17}$$

$$\frac{d}{dt}y = \frac{y^2}{2x} - \frac{1}{2}x. \tag{3.18}$$

From the above, the solutions of the equations will be the family of transformed circles, all sharing a common tangent point at the origin. Such an intersection of many phase space trajectories is not unusual. An attracting fixed point, for example, is approached asymptotically for all initial conditions in its basin of attraction (i.e., solutions are unique for finite times). What is unusual about these equations is that the common point is intersected in finite time, and further is not a fixed point. This is easily seen by taking the limit of the equations along a solution of radius r:

$$\lim_{x,y \to 0} \frac{d}{dt}x = 0, \tag{3.19}$$

$$\lim_{x,y \to 0} \frac{d}{dt}y = r. \tag{3.20}$$

Thus, the origin is a singularity of the equations, where neither past nor future time evolution is uniquely determined. Consequently, we refer to these equations as the nondeterministic harmonic oscillator (NDHO).

The NDHO provides a simple example of the type of system we are examining: solutions of the equations of motion are a family of closed loops ("transients"—since they are all divorced from their "past") all sharing a common tangent point. From the limit equations, the dynamics of the NDHO are not defined by the equations of motion alone (this is not a necessary condition for a system to be nondeterministic). However, suppose we built an NDHO in a laboratory. How would it behave? We note that all physical systems are subject to external perturbations, or "noise." While the physical state of our NDHO is far (in phase space) from the point (0,0), external noise will have little effect, provided the average amplitude of the fluctuations is small compared to r for that trajectory. However, as the trajectory approaches the origin, noise plays a larger role. Solutions for all r converge together, ultimately intersecting at (0,0). Thus, noise will cause the trajectory to jump between solutions of widely differing r in a random way.

What is the effect of this on our laboratory NDHO? Suppose we begin the system on a solution of radius r_1. As we watch the system evolve forward in time, we will find that after it passes near the origin the trajectory has changed to a completely different solution of radius r_2. Repeating the experiment with the same initial conditions, we find that the trajectory jumps to a completely different solution of radius r_3, where $r_3 \neq r_2$. Were we to repeat this a large number of times, for different values of r_1, we would find that the solution after the singularity is completely unrelated to the solution before. If the NDHO were allowed to run for several oscillations, a time series measurement of one variable would appear as a piecewise continuous sequence of oscillations with different amplitudes. Furthermore, the sequence of amplitudes would be random and unpredictable. We term this behavior nondeterministic chaos, nondeterministic because its origin lies in the nondeterminism at a non-Lipschitz singularity, and chaos because of the long term unpredictability of the dynamics. We note however that in between the jumps, the dynamics is fully deterministic

3.3.2 A Physically Motivated Example

The NDHO, while illustrative of the type of nondeterminism we are examining, is also a somewhat contrived example. We now describe a nondeterministic system based on physical considerations. This system is a model of the behavior of neutron star magnetic fields. We describe it briefly:

The model envisions two oppositely charged spherical shells which are allowed to rotate differentially. The magnetization of one shell ($\mathbf{M}_1$) will interact with the magnetic field of the second shell ($\mathbf{H}_2$), as well as experience non-electromagnetic ("mechanical") interactions with the surrounding medium. The magnetic interactions include a term to induce precession of ($\mathbf{M}_1$) about the instantaneous direction of ($\mathbf{H}_2$), and the Landau-Lifshitz magnetic damping, which tends to align ($\mathbf{M}_1$) with the direction of ($\mathbf{H}_2$). The mechanical interaction is taken as a simple damping, proportional to the difference in angular velocities of the two spheres. Parameterizing the interactions, we obtain the following equations:

$$\frac{d}{dt}\mathbf{M}_1 = \bar{\gamma}(\mathbf{M}_1 \times \mathbf{H}_2) - \bar{\lambda}(\frac{\mathbf{M}_1 \cdot \mathbf{H}_2}{\mathbf{M}_1^2}\mathbf{M}_1 - \mathbf{H}_2) - \bar{\eta} \cdot (\omega_1 - \omega_2). \quad (3.21)$$

Following a scaling procedure, and conserving angular momentum, we arrive at the following equation:

$$\frac{d}{d\tau}\mathbf{m} = -\mathbf{m} \times \hat{z} - \lambda(\frac{\mathbf{m}\cdot\hat{z}}{m^2}\mathbf{m} - \hat{z}) - \bar{\bar{\varepsilon}}(\mathbf{m} - \hat{z}), \quad (3.22)$$

where $\mathbf{m}$ is the scaled magnetization, τ is the scaled time, λ is the scaled Landau damping parameter, and $\bar{\bar{\varepsilon}}$ is the scaled viscous damping parameter tensor.

Examination of the equation reveals axial symmetry about the z-axis. This prompts us to make the following transformation:

$$x = \sqrt{m_x^2 + m_y^2}, \quad (3.23)$$

$$z = m_z, \quad (3.24)$$

$$\phi = \arctan\frac{m_y}{m_x}, \quad (3.25)$$

which implies

$$m_x \to x\cos\phi, \quad (3.26)$$

$m_y \to x\sin\phi$, (3.27)

$m_z \to z$. (3.28)

Substituting these transformations into the scaled equation

$$\dot{x} = \frac{\lambda xz}{x^2 + z^2} - \varepsilon x,$$ (3.29)

$$\dot{z} = \frac{\lambda z^2}{x^2 + z^2} - \bar{\varepsilon}z - (\lambda - \bar{\varepsilon}),$$ (3.30)

$$\dot{\phi} = 1.$$ (3.31)

where an overdot again represents differentiation with respect to the scaled time τ. The φ equation is trivial, simply representing a constant precession about the z axis.

At this point, the reader may be concerned about the seemingly unphysical nature of intersecting phase space trajectories (Figs. 3.8, 3.9). However, we are discussing the behavior only in the presence of noise. The singularity, is never actually encountered as revealed by probability theory. Even so, the nature of the solutions near the singularity combined with the presence of random perturbations will have a definite effect on the dynamical behavior. Another interesting technical point is that in the absence of noise, the equations are actually deterministic. It can be shown that if the singularity is approached along any of the solutions of the equations, their derivatives are uniquely defined. This is to be contrasted with the NDHO. These equations are thus not rigorously nondeterministic. However, the introduction of any noise, no matter how small, destroys the determinism of the neutron star model, and so in any physical situation, we deem the equations of motion to be effectively nondeterministic at the singularity. Of course, noise itself is stochastic, so rigorously speaking, the preceding statement applies to all dynamical systems. As we shall see in the following section, what one really should examine is the large scale time-averaged behavior in the presence of noise compared to the classically ideal, zero

noise behavior

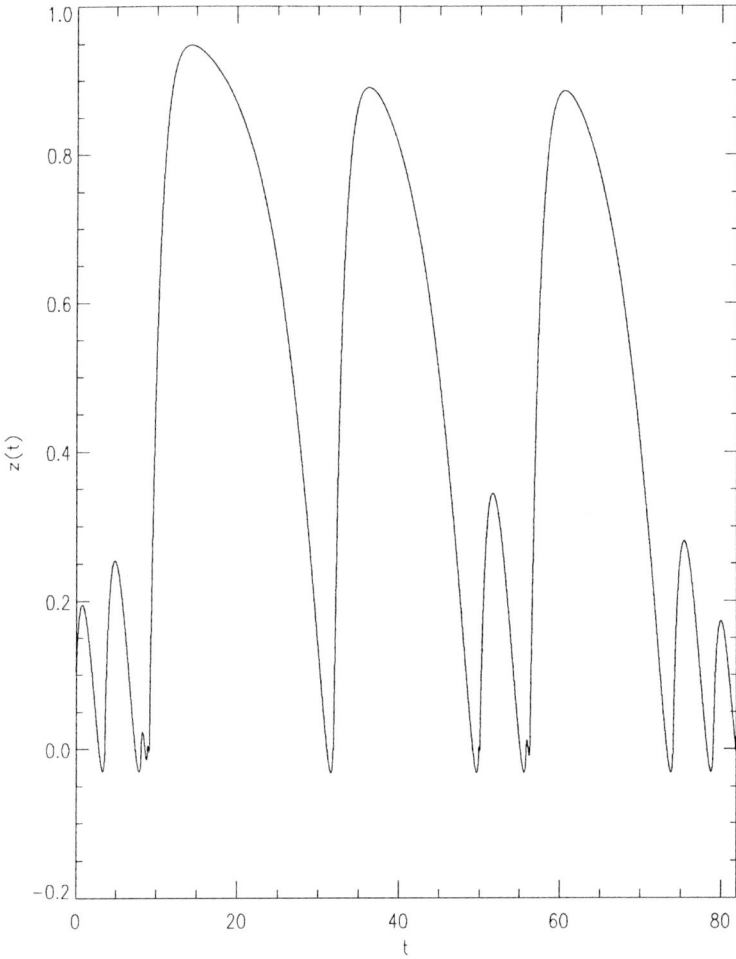

Figure 3.8. Time series of neutron star equations. Note there are many oscillations not easily visible. Compare with the following phase plane plot, and see Fig. 3.14 below for additional detail.

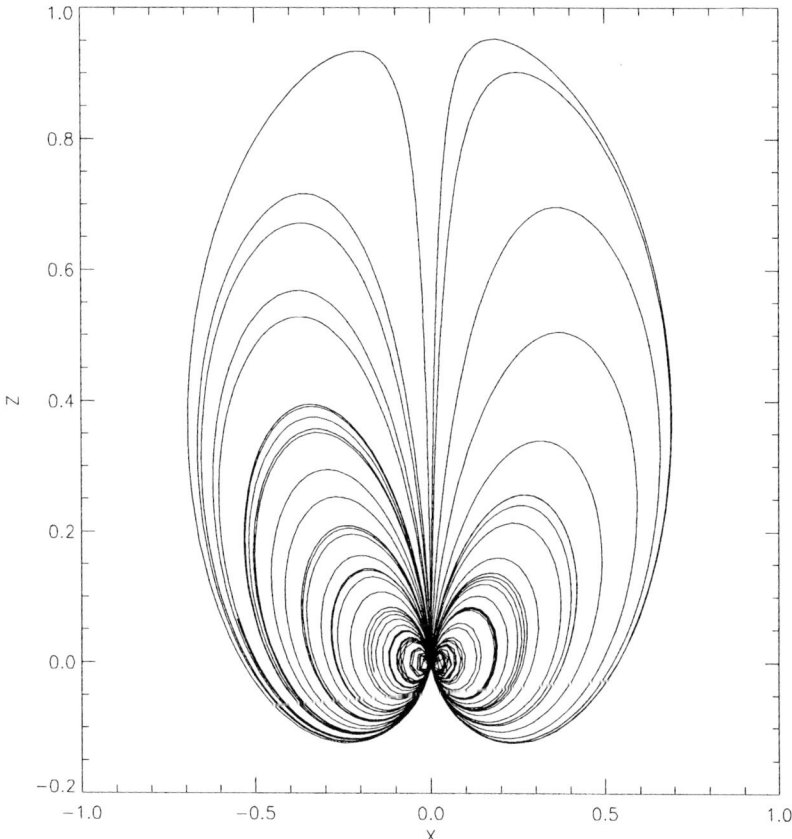

Figure 3.9. Phase plane plot for the neutron star equation.

3.3.3 Uncertainty in Piecewise Deterministic Dynamics

Let us now take a close look at the effect of uncertainty on a dynamical system in an effort to see what is implied by the existence of the type of nondeterministic singularity we have described. While one often describes systems in the classical domain by deterministic rules, a real system is always subject to uncertainty due to inaccuracy in measurements, the effect of external perturbations, and ultimately quantum mechanical uncertainty. Rather than attempt to account for all of these via one single equation, we divide an experiment into "system" (the thing being studied) and "noise" (everything else).

Unstable Singularities

As the "noise" generally involves many (10^{26}) degrees of freedom, and is often attributable to causes outside of our control and knowledge, we are forced to treat the noise in a statistical manner.

In general, we think of a dynamical system as a set of differential equations

$$\dot{x} = f(x), \qquad (3.32)$$

which describe the evolutions of the J dependent variables x_j in some J-dimensional phase space (we will confine our discussion to autonomous systems; i.e., time-independent). In purely mathematical terms, each point of the phase space is unique and distinguishable. For a physical system, however, this is not true. The noise implies a minimum uncertainty, which we shall call δ. Two points closer that δ will be indistinguishable. Let us then divide up our phase space into many regions of size δ each with a volume $\propto \delta^J$. We may then think of calculating the average behavior of the general dynamical system in each volume, as well as the expected variance about this average.

Suppose that $f(x)$ is a polynomial, or Taylor expandable within a given volume centered on a point x_0 (all such functions necessarily satisfy the Uniqueness Theorem). For notational simplicity, we take our example as the two-dimensional system, though the result applies to systems of any dimension. The Taylor series in this case is

$$f(x_0 + x, y_0 + y) = \sum_{n=0}^{\infty} \frac{1}{n!} (x \frac{\partial}{\partial x} + y \frac{\partial}{\partial y})^n f(x,y) \Big|_{x=x_0, y=y_0}$$

$$= a_{00} + a_{10}x + a_{01}y + a_{20}x^2 + a_{02}x^2 + a_{11}y + \ldots, \qquad (3.33)$$

where a_{ij} represents the series coefficients. The average value of $f(x,y)$ in a given volume is

$$\langle f(x_0 + x, y_0 + y) \rangle \frac{1}{V} \int_{-\partial/2}^{+\partial/2} \int_{-\partial/2}^{+\partial/2} (f(x_0 + x, y_0 + y) dx dy$$

$$= \int_{-\partial/2-\partial/2}^{+\partial/2+\partial/2} \int dxdy = \partial^2, \qquad (3.34)$$

where we have chosen the region as a cube of side δ centered on (x_0, y_0) with total volume $V = \delta^2$. Applying this to the Taylor series

$$\langle f(x_0 + x, y_0 + y) \rangle = a_{00} + \frac{1}{3}\partial^2 (a_{20} + a_{02} + \dots) \qquad (3.35)$$

Note that terms with odd powers in x or y cancel when integrated over the symmetric interval, thus the average value $\langle f(x_0+x, y_0+y) \rangle$ is simply $f(x_0, y_0)$ plus correction terms in even powers of δ. Assuming $\delta \ll 1$, we keep only the leading correction term.

Having found the average value of $f(x,y)$ in a cell, we now wish to find the uncertainty in this value, and especially how this uncertainty relates to the fundamental uncertainty scale δ in phase space. The root mean square (RMS) of $f(x,y)$ over a cell is simply

$$\sigma = (\langle f(x_0+x, y_0+y)^2 \rangle - \langle f(x_0+x, y_0+y) \rangle^2)^{1/2},$$

$$= \delta(\frac{1}{3}(a_{10}^2 + a_{01}^2))^{1/2}. \qquad (3.36)$$

Not surprisingly, we find that for "nice" functions, the uncertainty in the dynamical vector field goes like the inherent uncertainty in phase space. Thus in the "classical" or "thermodynamic" limit, where δ is taken as very small, we find that the dynamics is essentially unchanged. The result applies even for deterministic chaos. The Lorenz equations, for example are polynomial in the phase space variables. In the chaotic parameter regime, the nonlinearity acts to spread any initial uncertainty across the strange attractor (see above). However, this is (loosely speaking) a global property of the system. Locally, the dynamical uncertainty is related only to the uncertainty in the phase space variables.

What happens when we apply the above analysis to the types of systems we are discussing? Both the NDHO and the neutron star model contain essential singularities, and hence cannot be Taylor expanded around these points.

However, we can explicitly calculate average value and variance of the equations of motion in a phase space cell about the singularity. (We emphasize here the need for statistics.) Consider the neutron star model. First, we note that because of the nature of the singularity, the quantities obtained for the various integrals will depend on the order of integration. This is easily dealt with by transforming to plane polar coordinates $(x,z) = (r \sin \theta, r \cos \theta)$, and taking the phase space cell as a disk centered on the singularity. Performing the integration over this cell, we find that the average values are well defined:

$$(<\dot{x}>, <\dot{z}>) = (0, \varepsilon - \frac{1}{2}). \tag{3.37}$$

The RMS deviations from these values are

$$\sigma_x = \sigma_z = \sqrt{\frac{1 + 2\varepsilon^2 \delta^2}{8}}, \tag{3.38}$$

and we now begin to see the fundamental difference between "smooth" deterministic and nondeterministic "piecewise" deterministic dynamics. In the classical limit, the uncertainties in the above equation do not become arbitrarily small. As $\delta \rightarrow 0$, we find that the uncertainty in the RMS of the transformed neutron star equations goes to $1/2\sqrt{2}$, which is of the same order as $(<\dot{x}>, <\dot{z}>)$. In the presence of any uncertainty, regardless of how small, the equations of motion do not even approximately determine the behavior near the singularity. This is why we term the dynamics here as nondeterministic.

Having established this key distinction, let us briefly compare deterministic chaos and piecewise deterministic dynamics. Again, the action of a deterministic chaotic system is to spread an initial uncertainty over a larger and larger portion of the attractor as time goes by. This spreading is caused by a global instability, associated with one or more positive Liapunov exponents. For piecewise deterministic dynamics, the uncertainty is essentially induced at a particular point, or more precisely, in some small localized region. The effect of this can be illustrated by using the information approach. Assuming that the amount of information contained in some region of phase space is proportional to the volume of the cell, the rate of information generation is

$$\frac{dH}{dt} = \frac{1}{V}\frac{dV}{dT}, \tag{3.39}$$

where the derivative of the system is defined as

$$\frac{1}{V}\frac{dV}{dt} = \sum \frac{\partial_i \dot{x}_i}{\partial x_i}. \tag{3.40}$$

The total change in the system information may be found as the integral of the derivative with respect to time along the system trajectory. For a deterministic chaotic system, information is created at a rate proportional to the largest positive Liapunov exponent. Noting that an increase in system information corresponds to a decrease in the knowledge of an observer, the phenomenon of deterministic chaos implies a steady decrease of an observer's knowledge of the system's past, i.e., its initial conditions. Another way of looking at this is to say that nearby points on a trajectory have a high degree of correlation, while this correlation decreases as the points become more separated. Indeed, this property may be utilized when attempting a time series reconstruction of a chaotic system.

For the piecewise deterministic case, we examine the NDHO, as we know the solutions analytically. Taking the initial conditions to be $(x,y) = (0,0)$ at $t = 0$, the solutions for the equations are

$$x = A(1 - \cos t), \tag{3.41}$$

$$y = A \sin t. \tag{3.42}$$

The reader may at this point be concerned that we give the initial conditions at the singularity. We justify this by noting that with regard to information calculations, we can only talk about those points which are distinguishable. Since there will always be some uncertainty in the system, we are really referring to some set of points near the origin, with the initial conditions being somewhere with this region. The derivative of the NDHO is given by

$$\frac{1}{V}\frac{dV}{dt} = \frac{y}{z}. \tag{3.43}$$

Integrating along a solution over half a cycle, we find the total information generated to be

$$\Delta H = \int_0^\pi \frac{\sin t}{1-\cos t} = \log(1-\cos t)|_0^\pi = \infty . \tag{3.44}$$

The gain in system information is infinite, which means that the observer has zero knowledge of the system's past. (Note the similarity with equilibrium nondeterministic systems.) This is another key feature of nondeterministic dynamics: whenever a trajectory passes near the singularity, the future time evolution is completely decoupled from the past.

3.3.4 Nondeterminism and Predictability

Nondeterministic chaos has the property of being predictable for short times (between intersections of the singularity), yet completely unpredictable over long time periods. Long-term unpredictability is also one of the hallmarks of deterministic chaos, but it is here that the similarity ends. Aside from being described by deterministic equations, deterministic chaos is often characterized by exponential divergence of initially close solutions, and associated with an complex fractal structure, the strange attractor. Nondeterministic chaos derives its unpredictability from a more violent, but localized instability. Further, there exists no attractor, strange or otherwise, at least in the usual sense of the word.

Formally, we can study the effect of noise on a "classical" system by constructing a Langevin equation

$$\dot{x} = f(x) + \varepsilon\phi(t), \tag{3.45}$$

where $f(x)$ represents the classical part of the equation of motion, $\varphi(t)$ is some random function with mean zero and standard deviation of one, which reflects the statistics of the noise, and ε controls the average size of the random perturbations. The trajectory in phase space is a Brownian motion, and as such there exists a time-dependent probability density of finding the system state at some particular point in phase space. Any initial probability density will tend to diffuse through phase space in a manner governed by the forward Kolmogorov equation. The rate of this diffusion is directly related to the rate at which the system generates information, which is a measure of the system's predictability.

For the systems we have considered, the classical part of the Langevin equation is singular, and so the associated forward Kolmogorov equation will also be singular. The analysis in the previous section indicates that we expect an explosion of information at the singularity, and this can be illustrated by integrating the Langevin equation for the neutron star model for some set initial conditions confined to a small region of phase space. For this simulation, we took $\varphi(t)$ to be distributed as a Gaussian, and ε approximately the time-step. Figure 3.10 illustrates the generation of information at the singularity due to the external noise. Initially, we see smooth deformation of the initial set of points. However, once the singularity is encountered, points are scattered, and soon are randomly spread through a region of phase space [in this case, the region is enclosed by the lines of the saddle at (0,1)]. This behavior is in stark contrast to what one expects from deterministic chaos, where an initial volume is stretched and folded, spreading across the attractor in a smooth fashion. Dynamical measures such as the Liapunov exponent are meaningless. The "attractor" exists only in a statistical sense, representing the probability density that a particular point in phase space will be visited. This distribution is shown in Figure 10, and was calculated by integrating the Langevin equation for 200 million time steps.

Figure 3.11 gives a long-term, global statistical picture. However, the nature of nondeterministic chaos allows us to easily extract more useful statistical information. In particular, we shall utilize the fact that away from the singular point, the dynamics is quite well-behaved. Let us return to the NDHO, as its solutions are known analytically. The solutions of the NDHO may be parameterized by their radius. Solutions away from the singularity have essentially constant radii; the big jumps occur only near the singularity. Can we predict the probability that a circle of given r is chosen when the orbit leaves some neighborhood about the origin? Let us define this neighborhood as a disk of radius δ, and note that an orbit leaving this neighborhood does so with an angle θ, which we take as measured from the y-axis. Now, the probability density of picking a particular θ is constant, i.e.

$$p(\theta)d\theta \propto d\theta. \qquad (3.46)$$

Next, we note that everywhere except at the origin the Existence and Uniqueness Theorem applies to solutions of the equations, thus each circle of radius r is associated with a unique θ, and we may write θ as a function of r. Substituting into $p(\theta)$, we find

$$p(r)dr \propto \frac{\partial \theta(r)}{\partial r} dr. \qquad (3.47)$$

The probability of getting a circle between r and $r + \Delta r$ is simply

$$P(r, \Delta r) \propto \int_{r}^{r+\Delta r} \frac{\partial \theta(r)}{\partial r} dr = \theta(r + \Delta r) - \theta(r). \qquad (3.48)$$

For the case here, we find

$$P(r, \Delta r) \propto \arccos \frac{\delta}{2(r + \Delta r)} - \arccos \frac{\delta}{2r}. \qquad (3.49)$$

This approach is somewhat simplified. A rigorous derivation would account for the statistical properties of the noise, and derive $P(r, \Delta r)$ via stochastic calculus. The above does show, however, that the simple structure of the solutions of a nondeterministic system lends itself to the construction of statistical arguments. Furthermore, with a judicious choice of δ, based on knowledge of the average amplitude of the fluctuations, the procedure should yield a good approximation of the true distribution $P(r, \Delta r)$.

3.3.5 Controlling Nondeterministic Chaos

The control of deterministic chaotic systems using small perturbations has been a subject of much research. The most popular method of controlling deterministic chaos involves the stabilization of (otherwise) unstable periodic orbits which are embedded in the chaotic motion. As

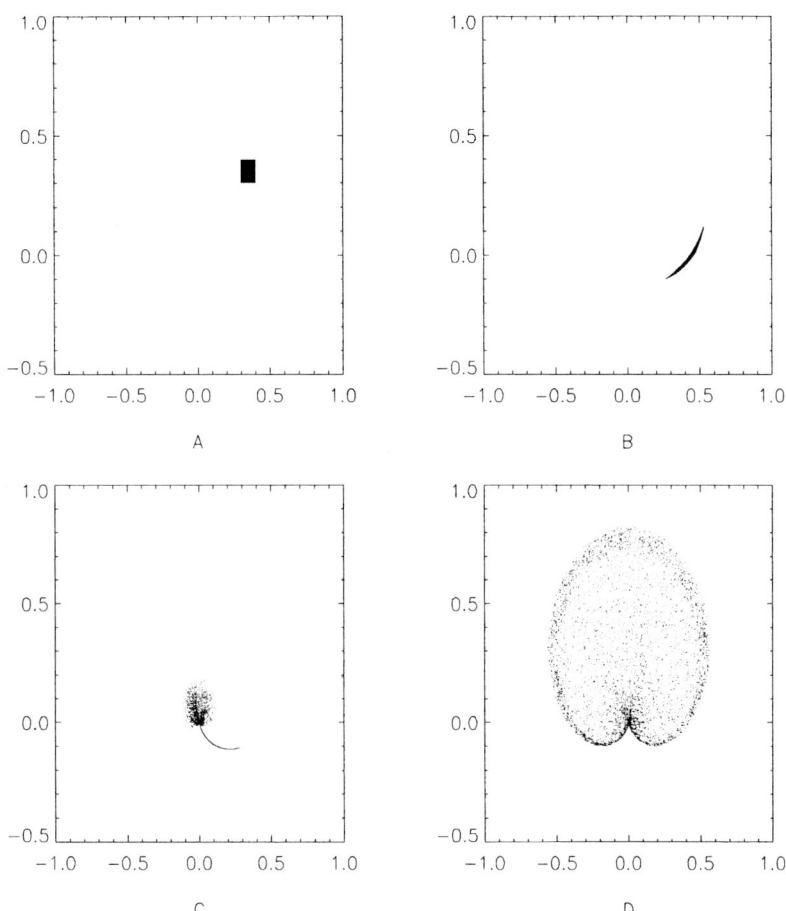

Figure 3.10. Loss of information due to the singularity. Clearly, in this situation, although there is a determinism, practically speaking, only stochastics can make it meaningfully understood. A) 10,000 initial points are arranged in a 100 x 100 square. B) There is an initial evolution. C) When the singularity is encountered, all points are randomly scattered. D) All information of the initial conditions is lost. The only information available is in the density of the trajectories. Thus knowledge of the system can only be gained through the probabilities.

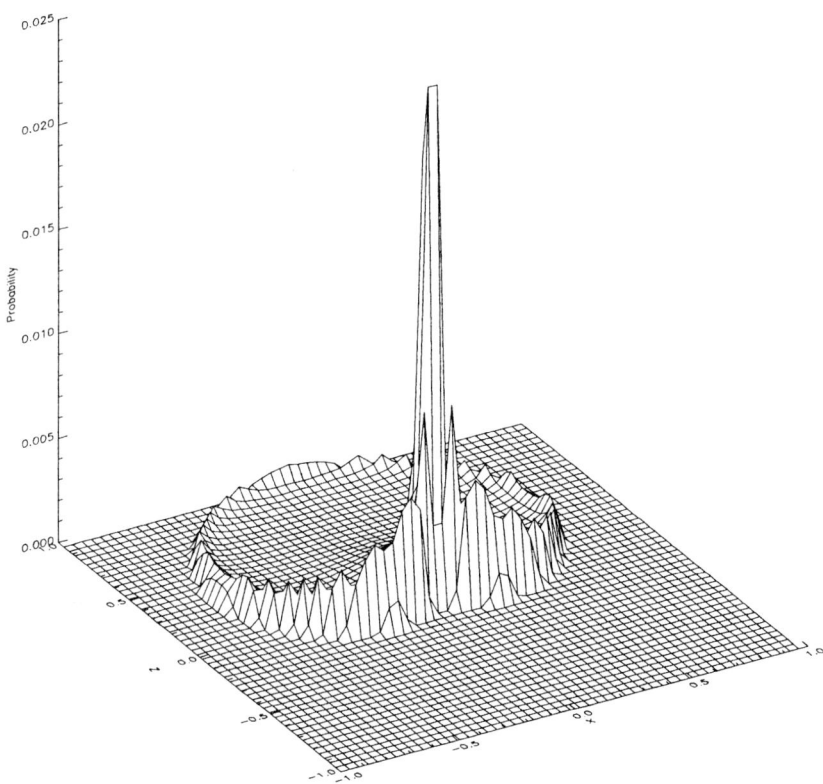

Figure 3.11. Probability of finding the system in a particular region of phase space. The distribution was found by integrating the equations for 200 million time steps and totaling the amount of time spent in each region.

there exist an infinity of orbits, a rich variety of behaviors may be extracted from the controlled deterministic chaotic system, allowing for flexibility and easy optimization of a system's behavior.

For a nondeterministic chaotic system, we have a similar situation. With a continuum of different solutions intersecting at a single point, we can easily effect control via an appropriate perturbation. Similar to the previous section, we simply examine how solutions leave a δ-neighborhood about the singularity. Again, away from the singular point the solution is well-defined. Suppose that each different solution may be parameterized by some quantity γ (in the case of the NDHO, this is the radius). A given solution, parameterized by γ_0, will intersect the δ-neighborhood at a unique point (x_0,y_0). From this, we may construct the angle $\theta(\gamma_0) = \arctan x_0/y_0$.

The angle $\theta(\gamma)$ we term the control angle, and the reason should be obvious. To keep the system on a solution with parameter γ_0, we need only to wait until the trajectory approaches the origin, and then perturb it so that it leaves at angle $\theta(\gamma_0)$. This perturbation will be quite small, of the order of δ with the size of δ being determined largely by the noise amplitude. We see that in a nondeterministic system, there is a continuum of possibilities available through small control perturbations. If a change in system behavior were required, it is easily and quickly effected by simply changing $\theta(\gamma)$. In fact, one could vary $\theta(\gamma)$ as a function of time to induce arbitrarily complex behavior.

As an example, we have applied this control algorithm. To simulate the effect of noise, a small (10^{-4} of the integration stepsize) normally distributed random number was added at each integration step. The controlled signals for various values of θ are shown in Fig. 3.12. Figure 3.13 shows the effect of noise for different values of δ.

3.3.6 Implications

The type of "nondeterminism" should not be considered in the same class as "stochastics." (Although some would argue that it should be.) At the very least it is clearly different from the "nondeterminism" which is the opposite of determinism. Indeed, the behavior of both the NDHO and neutron star model are uniquely determined away from the singular point. It is at this point, and this point only, that the nondeterministic nature of the equations arises. In the presence of random fluctuations, which are ubiquitous (though perhaps small) in physical systems, the nondeterminism, albeit it at a single point, becomes important. The resulting dynamics, which we have termed nondeterministic chaos, consist of a random sequence of "transient" oscillations, which are strictly speaking divorced from each other, yet they are part of the same process. A difficulty in appreciating this view may be the implied psychology. The classical approach has fostered the perception of dynamics which are smooth and continuous; whereas the present exposition suggests an alternative perspective based on discontinuous processes. Certainly, the argument may be that human perception itself counters this view. Yet it may be pointed out that there are several common experiences which suggest that there is really nothing that needs to be dissonant. Specifically, consider movies and animations which, as is well known, are created from many independent frames which produce the so-called illusion of motion. Clearly perception, as psychologists inform us, is at the mercy of our physiology, which takes time to process—the time between seeing something and apprehending it is not instantaneous.

3.4 *Classification of Nondeterministic Systems*

There are other examples of such nondeterministic systems. Chen has independently suggested the same behavior under the heading of "noise induced instability" (though nondeterminism as such is never explicitly mentioned). For clarity, equilibrium-based mechanics may be called "Type I non-determinism"; whereas those not dependent upon equilibrium as "Type II non-determinism."

The primary difference between the two is that for Type I nondeterminism, the singularity occurs at an equilibrium point of the equations of motion, while for Type II, the singularity is shared among a group of dynamic trajectories. The physical implications of this difference are yet to be explored.

Whether as system is nondeterministic as here presented can be an important question. As has been seen, issues of prediction and control would be addressed much differently for a nondeterministic system as opposed to the more traditional "deterministic" chaos. Indeed, Crutchfield has shown that in the context of model building, assuming determinism when the underlying process is nondeterministic leads to undue complexity in the model—very similar to the "over-determined" processes of statistics. It would seem reasonable to search for nondeterministic chaos in apparently complex systems, especially in cases where traditional analysis tools (which, again, assume determinism) have failed. But again, before even this, care must be obtained in understanding the system, as well as the methods used to record the system behavior (see Mathematical Appendix).

This is a point which is sometimes easily overlooked, since many of the data acquisition methods currently developed for computer routines make it so easy to record observables. In all these instances, however, it should be understood that typical analogue-to-digital recording assumes a linear systems theory approach. Although this may be obvious to some, there are many biological scientists who have not been made aware of this somewhat subtle point. This issue becomes even more crucial when models of phenomena are attempted. Models may look quite good, but the question may be what is being modeled—the data or the data as it was acquired. This is more than simply a question of noise artifact.

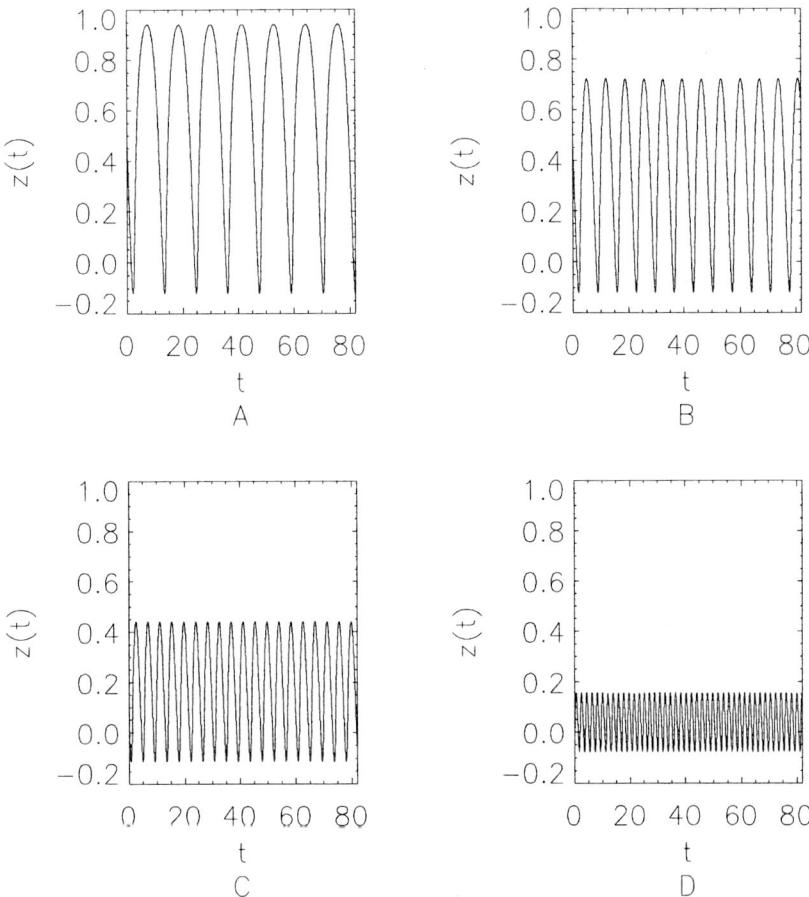

Figure 3.12. Example of neutron star model when the control algorithm is applied. Signals are shown for $\theta(\gamma) = $ a) 0.005, b) 0.03, c) 0.08, d) 0.2.

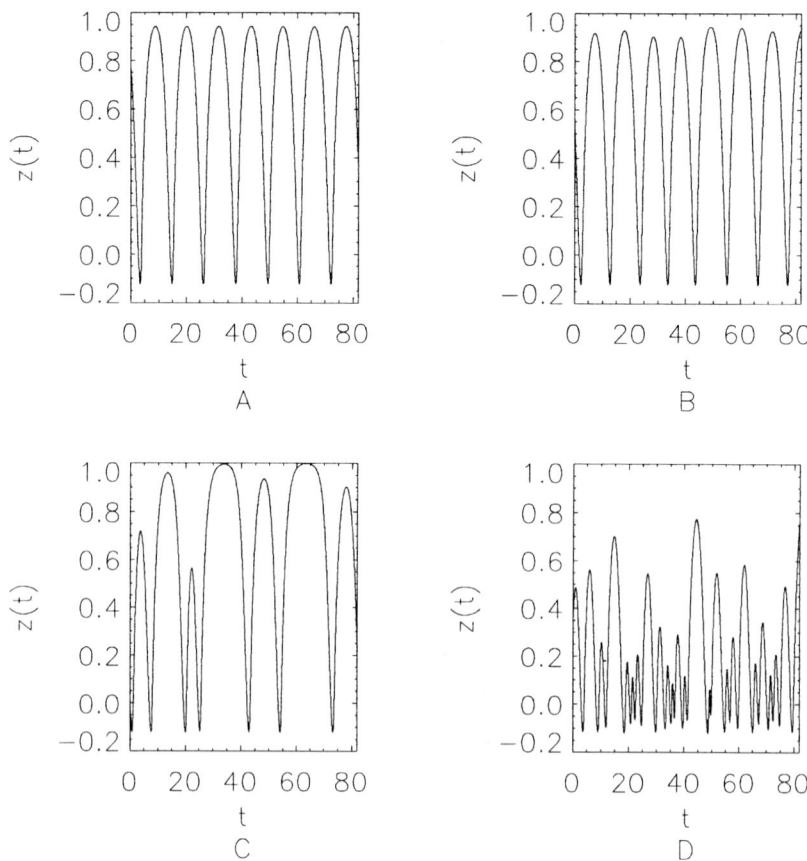

Figure 3.13. Control breaks down if δ is chosen close to the noise level. Signals are chosen for $\delta =$ a) $10^4 \sigma$, b) $10^3 \sigma$, c) $10^2 \sigma$ and d) 10σ., where σ is the RMS noise.

I stress the point that contemporary signal acquisition techniques are based on analogue-to-digital conversion. The subject of frequency of sampling is little discussed (except, perhaps by compact disc music engineers who have done much work on an ad hoc basis). The traditional approach is based on the so-called Nyquist frequency, which is based on, a has been emphasized, the ubiquitous linear systems theory. If a signal contains nonlinearities or singularities, slavish application of its tenets can lead to false conclusions (Fig. 3.14).

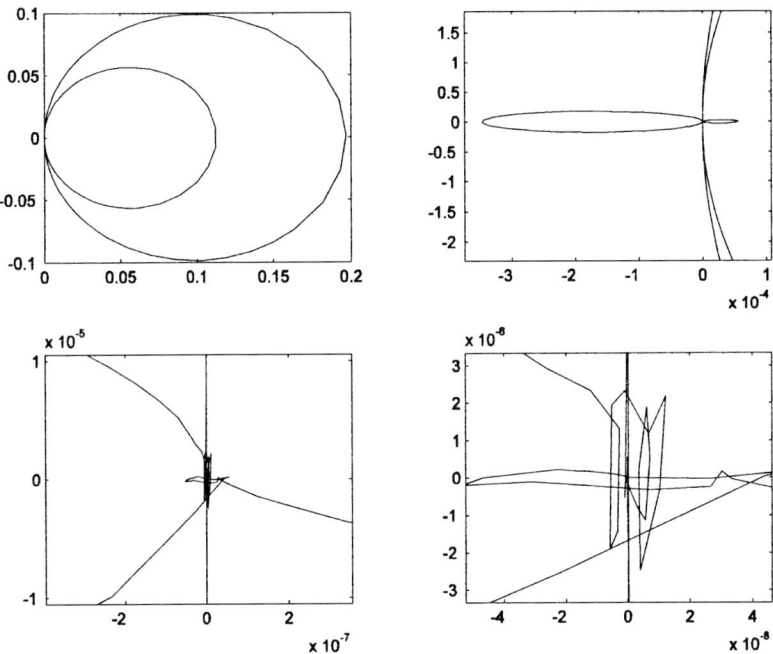

Figure 3.14. Progressive details of the SHO—note the circles not seen in the top left.

4 *Singularities in Biological Sciences*

A basic tenet in biology and medicine of the last twenty years has been that biochemical and physiological variables are controlled within narrow limits by feedback processes in homeostasis dynamics. The feedback loops that enable homeostasis have also been described mathematically, and this result has been considered a valid support to the thesis that mathematics has a substantial role in quantifying biological processes. These mathematical models revealed a feature of biochemical and physiological observables that appeared to be important: considering changes to the delay incorporated into a feedback loop, or to the outputs of the components of the loop, it was possible to obtain regular oscillations about some mean value. Similar mechanisms were assumed to be responsible for observed rhythmicities with periods ranging from under a second as electrical rhythms in the brain, through minutes and hours, to a day or a year. Other patterns were recognized such as, in particular, bursts of activity in neurons.

Some two decades ago, a new branch of biochemical and physiological dynamics was developed using the tools of nonlinear functions. Nonlinear equations are more complex mathematical objects than ones describing a straight line, and the output value from one calculation for such functions becomes usually the input value for the next and so on, going ad infinitum. Such mathematical features were considered to exhibit an evident connection with the continuous operation of a feedback loop.

The employment of such new mathematical techniques produced results that at the first appeared counter-intuitive but opened the way to the so called analysis of chaotic dynamics of biochemical and physiological variables. If the feedback from the output into the subsequent input is made more complex or if the mathematical function describing the relationship between input and output of any element in the feedback loop is changed, then different and far more complex results can ensue.

Nonlinear system theory became widely used in recent years to characterize the behavior of a physiological dynamical system from a single experimental time series especially in the analysis of electrocardiograms (ECG) and electroencephalograms (EEG). The concept of the cardiac rhythm as an expression of a periodic oscillator was challenged in the late when research showed the heart to be associated with irregular and possibly chaotic deterministic dynamics. Studies indicated that externally stimulated cardiac tissue develops patterns that are characteristic of nonlinear dynamical systems. Some authors advanced the suggestion that the fractal structure of the His-Purkinje system should represent a structural substrate of chaotic-deterministic cardiac dynamics.

It was deduced that strictly periodic cardiac dynamics should not represent a healthy condition but, on the contrary, turned out to be correlated with pathological states. The whole paradigmatic framework of biology and medicine was designed for change.

Since then, the idea that nonlinear methods might reflect properties associated with normal as well as pathological heart functioning more accurately than conventional ones, stimulated increasing interest in the application of tools from nonlinear dynamics to a multitude of biological and physiological signals. Considerable interest has been devoted to applying deterministic chaotic nonlinear analysis to the dynamics of biological systems. Currently, some physiological and pathological systems seem to exhibit what could be called apparent random behavior whose actual nature must be further investigated. Other physiological systems seem to exhibit chaotic-deterministic dynamics, however, the final determination is open. One difficulty is that the terms and methods used sometimes lack the desired level of rigor.

These studies often employ the calculation of dynamical invariants, such as correlation dimension, Liapunov exponents, and entropy to determine the existence of deterministic-chaos. The basic difficulty remains regarding the determination as to whether the experimental time series is generated by a deterministic chaotic dynamics or by an alternative mechanism having strong stochastic features but arising from non deterministic chaotic behavior. It has been demonstrated that these usual measures, such as saturation of correlation dimension and existence of positive Liapunov exponents cannot by themselves establish the existence of deterministic chaos. Furthermore, improper application of these methods question the results.

Errors associated with the acquisition of data such as inappropriate sampling frequency, noise filtering, limited stationarity of the signal, and digitization errors can lead to uncertainties in the value of the correlation dimension. For a time series without noise, the largest Liapunov exponent, λ_{max}, gives the exponential rate of divergence of two neighboring trajectories in the phase space. However, the existence of positive λ_{max} is true also for stochastic dynamical systems. Consequently the existence of a positive exponent does not necessarily indicate that a given system is deterministically chaotic.

The point to be made is that although systems may exhibit stochastic and deterministic features; this in itself is not sufficient to consider deterministic chaos the only possible explanation. In this sense, the elaboration of deterministic chaos has become a double-edged sword: it demonstrated the possibility that natural phenomena can speciously appear stochastic, but at the same time it limited the investigative focus to this narrow domain.

4.1 An Alternative Approach

In the previous section it was pointed out that many physical, biological or physiological variables exhibit apparently random behavior. Some difficulties remain regarding the determination that they are truly chaotic. In fact, some fundamental questions need to be explored.

The first point concerns the problem of establishing the real features of existing determinism in the dynamics of systems. Determinism has represented the basic paradigm of physical science for at least 300 years. It may be defined as the proposition that each event is necessarily and uniquely a consequence of past (future) events: such events evolve in an ordered sequence directed by some "rule" fixing each successive (past) event on the basis of the preceding (future) events.

Determinism in itself was not a new concept developed by modern science. Certainly, the notion of "causality" could be traced through medieval scholastic philosophers all the way back to Aristotle. What was new was the removal of logic as the sole arbiter of this causality. Instead, in its place, observation of the material world attended by experimentation to develop prediction became the new arbiter.

The major feature of determinism is its rigidity. Chaotic determinism exhibits the same rigidity since, in spite of the chaotic behavior of the system, the strong requirement of the acting deterministic rule, remains. In deterministic systems there is a strong dependence of the system on its initial conditions. The deterministic system is linked to its initial conditions: at any time it operates only if it must remember the initial conditions from which it started.

In living matter especially it is difficult to accept such an extreme dependence of a system on its initial conditions. A physiological system, in fact, is required to adapt continuously its behavior to the requirements imposed from the environmental conditions in which it operates. This adaptability would seem to act at all temporal scales. At a qualitative level of elaboration, this is the first reason to have doubts regarding the possibility of applying a rigid deterministic paradigm to living matter.

The second reason regards more experimental facts which may be directly observed. Often biological systems exhibit oscillations in their behavior. These oscillations, however, are not always continuous. Instead, at times, biological signals exhibit "pauses" of varying length which beg for an explanation. Such pauses indicate, at the very least, nonstationarity.

To understand pauses one must reconsider determinism. Determinism of dynamical systems was admitted in science implicitly from the theory of differential equations in mathematics. Based on the classical (Laplacian) admittance of determinacy, the mathematical formalism of Lipschitz conditions, requiring boundedness of derivatives. The implication was one of a grand universe acting as some machine where all its parts uniquely related to all the other parts. All the motions of a system are uniquely determined starting with their initial positions and with initial velocities.

In determinism, the key is connected to the Lipschitz conditions whose validity assure uniqueness of solutions for differential equations and thus uniqueness of a trajectory for the dynamics of the corresponding system. As previously outlined, there appears to be a discrepancy between the universality of Lipschitz conditions on the one hand, and real behavior of physiological data. Going from a general, qualitative view to a more specific one, a possible interpretation of "pauses," are expressions of singularities arising in the mathematical counterpart describing the physiological phenomena. In other terms, in living matter the deterministic paradigm may fail, being delineated instead by a nondeterministic dynamics for systems in which the initial conditions no longer determine uniquely the state of the system at all later points in time.

The problem can be sketched with some refinement. The standard approach to physical and biological reality has assumed in the last 300 years a Laplacian point of view where it is accepted that the time evolution of a system is rigid, and it is uniquely determined by its initial conditions. The dynamics of the system remains linked to such initial conditions for the whole time of its evolution. This view of determinism is supported from some results of the mathematical branch of differential equations that are commonly considered as the basic mathematical framework of the determinism. In such theory, some existence and uniqueness theorems of differential equations assume uniqueness of solutions. (I stress the fact that this is an assumption—there is nothing inherent about such model equations that require such an assumption.) This is to say there exists uniqueness of time evolution in the dynamics of the system for which differential equations are admitted.

Such theorems require equations of motion of the system to satisfy Lipschitz conditions. However the deterministic approach is too rigid for biological dynamics where instead, a continuous adjustment of the system is required in order for the system to respond to its environmental requirements. In addition, actually, in the observation of recorded physiological data it is evidenced that often pauses appear in the data and thus such pauses suggest the non-universal character of the determinism and, indeed, the presence of singularities in the mathematical counterpart describing the dynamics. In fact the universal validity of Lipschitz conditions, assuming uniqueness of solutions for the dynamics of the system, was universally "admitted" in physical as well as in biological science. In other words, it is tacitly assumed that nature in its classical behavior is deterministic, and that, as consequence, the equations of motion, describing physical and biological systems, are Lipschitz.

There is no a priori reason to dogmatically admit that nature is universally Lipschitizian. The main concern here is a particular implication of non-Lipschitz equations. These equations admit non-unique solutions. If a dynamical system is non-Lipschitz at a singular point, it is possible that several solutions will intersect at this point. Since the singularity is a common point among many trajectories, the dynamics of the system, after the singular point is intersected, is not in any way determined by the dynamics before, and thus, in conclusion, this system will exhibit a dynamics that may no longer be considered to be deterministic (in the classical sense). An immediate consequence of such nondeterministic dynamics is the possibility of nondeterministic chaos. For a nondeterministic system, as the various solutions move away from the singularity, they will evolve very differently and will tend to diverge. Several solutions will coincide at the non-Lipschitz singularity, and therefore whenever a phase space trajectory comes near this point, any arbitrary small perturbation will put the trajectory on a completely different solution. As noise is intrinsic to any physical as well as physiological system, the time evolution of a nondeterministic dynamical system will consist of a series of transient trajectories with a new one chosen randomly (within the constraints of the observed probability distribution, which is constrained by the attendant physical/biological properties) whenever the solution nears the non-Lipschitz point in the presence of noise. In conclusion, such systems will exhibit what we call nondeterministic chaos.

Control mechanisms will act also in this framework as the previous chapter indicated. It is well known that the control of deterministic chaotic systems using small perturbations has been the subject of various studies. Control of deterministic chaos involves the stabilization of "unstable periodic orbits" which are embedded in phase space of chaotic motion. As there exist an infinity of orbits, a great variety of behaviors may be extracted from the controlled deterministic chaotic system. This allows optimization of system's behavior. In the case of systems exhibiting nondeterministic chaos, we have a similar situation, although it is stressed that the "orbits" here are not of the same type as the unstable periodic orbits of deterministic chaos: the orbits are part of a nondeterministic system—the orbits are not defined except through the observed probabilities. Similarly, "bifurcations" do not exist since there are no defined parameters of a deterministic system.

We have a continuum of different solutions intersecting at a single point. Control may be realized in this case via an appropriate perturbation. Similar to the case of control for deterministic chaos, we will have the possibility of optimization of the behavior for the system in the proximity of the singularity. Away from the singularity the solution is well defined. Suppose that each different solution may be parameterized by some quantity whose fixed value will be γ_0. A given solution will be parameterized by γ_0 and it will intersect the δ-neighborhood at a unique point (x_0, y_0). In conclusion, in nondeterministic chaotic systems, there are a continuum of possibilities available through small control perturbations.

The above arguments may be considered sufficient for an exposition of the alternative view. What remains is an outline of the importance of such a theory for application to experimental data. Nondeterminism and nondeterministic chaos could be a basic mechanism of many systems. This novel approach changes in a radical manner the paradigmatic view of real mechanisms acting in living matter.

Nondeterministic dynamics should be especially important in the case of biological data which have a close affinity with physical models such as oscillations. As an infinite number of trajectories would be accessible by an arbitrary small perturbation in the neighborhood of the singularity, selection of a particular solution by a control mechanism should be easily realized. Evidence of nondeterminism in biological systems is presented in the following section.

In conclusion the new theory should have its basic importance in explaining control mechanisms in living matter. This section was started by considering the particular role that was attributed to feedback in homeostatic biological systems. The traditional approach to physiological regulation has been one of negative feedback loops which are an expression of the principle of cause and effect and of determinism that instead, as has been suggested, would be violated in some specific and important moments of biological dynamics. In many cases, experimental evidence has been found that adjustments of biological and physiological variables are obtained in living systems before the supposed controller had the possibility to experience the feedback. The nondeterministic theory, here discussed, would be the model for such control. The examples are mainly concerned with the problem of detecting and analyzing physiological singularities.

4.2 Nonstationary Features of the Cardio-Pulmonary System

As is well known, the essential physiological functions of the cardio-pulmonary system are to provide gas exchange and to supply organs and cells with oxygenated blood. Alveoli provide exchange of oxygen and carbon dioxide in lungs. Fresh air flow into the lung through the airways constituting a 3-dimensional branching structure and diffuses through the thin walls of the capillaries in the blood. The oxygenated blood then enters the heart through the pulmonary circulation. The coordinated electrical activity of the heart provides a rhythmic contraction of the heart muscles and the oxygenated blood is pumped through the arteries in the body.

The experimental evidence is that the structure as well as the functioning of cardio-pulmonary system is complex. It includes many subsystems that are themselves inhomogeneous and irregular. The reopening of the closed airway segments during inspiration occurs in avalanches and the distribution of avalanches seems to follow a power law. The pulmonary vascular tree running parallel to the airway tree is a fractal structure and the resistance to blood flow of the tree shows scaling behavior.

The conduction system of the heart, through which occurs the propagation of voltage pulses, seems to generate complex patterns with fractal properties. Various physiological time series measured on the cardiopulmonary system appear to be extremely inhomogeneous and nonstationary. They all exhibit fluctuations according to an irregular and complex manner. Such noise, often neglected in studies on cardio-pulmonary system, carry instead an important role for function and in the structure of heart and lung systems.

Newborns and premature infants often develop irregular breathing patterns. Respiratory regulation develops substantial differences in the post natal period with respect to later life. The major underlying source of irregularities may be identified in brain stem rhythm generators or in immature central and peripheral chemoreceptors.

Generally speaking, several features may explain the systematic presence of irregularities measured in signals of the cardio-respiratory system. One is the existence of an inherent noise. Noisy operation of neurons, or heterogeneity of maturation of the vagal nerve are found to have a role for infants. Since myelination mainly determines the spread of propagation of action potentials, noise should appear as due to the heterogeneity of transmission times in a nerve made by a bundle of parallel neurons. The second is the observed presence of a singularity that produces large variations in inter-breath intervals.

Mathematical simulations of respiratory patterns have drawn extensively upon continuous variable models. However, these limit cycles systems are fully deterministic. They do not have the ability to reproduce all known experimental observations. In particular, they ignore the possible existence of nondeterministic components that instead could have a control role in living systems as has been outlined in the previous section. A hypothesis is that inspiratory and expiratory phases represent an automatic, or that is to say, a deterministic process of regulation but that some existing pauses (singularities) should insure that the appropriate automatic trajectory is selected during each and every breathing cycle.

Since these pauses insure that the appropriate automatic trajectory is selected for each and every breathing cycle, and since the whole system is geared to function properly under different physiological states and varied environmental challenges, it follows that this type of regulation strategy is critical since it permits cycle-by-cycle updates. As a consequence, the failure of this update mechanism may be implicated in a variety of pathological patterned breathing processes. Pleural pressure tracings of anesthetized rats, demonstrate kinks (see below) in the phase plots indicative of pauses. Their presence provide a more rational approach to the control of mechanically ventilated patients. The observation is made that the benefit, obtained by the so-called synchronized mode of ventilation, which initiates a breath near the inception of a spontaneous breath by the patient, is due to the presence of a nearby singularity.

4.2.1 Tracheal Pressures

Detection of non-Lipschitz dynamics in the respiratory system is demonstrated in Fig. 4.1 for intratracheal pressures which were recorded in the conscious rat using a special transducer-based technique. The intratracheal pressure trace was digitized at 500 Hz. Representation of this physiological variable in phase space inscribes rotational trajectories. Lack of superposition of the individual trajectories indicates they are of the branching type. This means that at the beginning of each and every cycle only one pressure trajectory of many can be opted. Non-Lipschitz dynamics in the plot is seen only during the various phases of the cycle during which pauses occur. The kinks (shown in a 3D plot at arrow) coincide with a ninety degree angle. It is in this region where the vectors are orthogonal, rendering the solutions to the operating differential equations of the non-Lipschitz type. This area actually constitutes a singularity for the system. With inspiration, each trajectory leaves the area from a slightly different position, approaches the various phases of every cycle and contracts back toward the orthogonal vectors. Each loop is unique. (Note the phases are indeed more complex for the tracheal pressures, but are beyond the scope of the present discussion.)

4.2.2 Lung Sounds

Using recurrence quantification analysis it was possible to detect singularities in nonstationary lung sounds recorded on both right and left apexes. This result indicates that respiratory dynamics is controlled by a discrete nondeterministic dynamics based on violation of Lipschitz conditions

Seven healthy subjects were recruited in this study. All subjects were non-smokers and without any known cardio-respiratory disease. Their mean age was 39 ± 14 years and their mean weight was 74 ± 17 kg.

Respiratory sounds were recorded on the left and right lung apexes. The subjects were asked to breathe spontaneously in the sitting position, according to the European Respiratory Society recommendations for short-term recording of breath sounds.

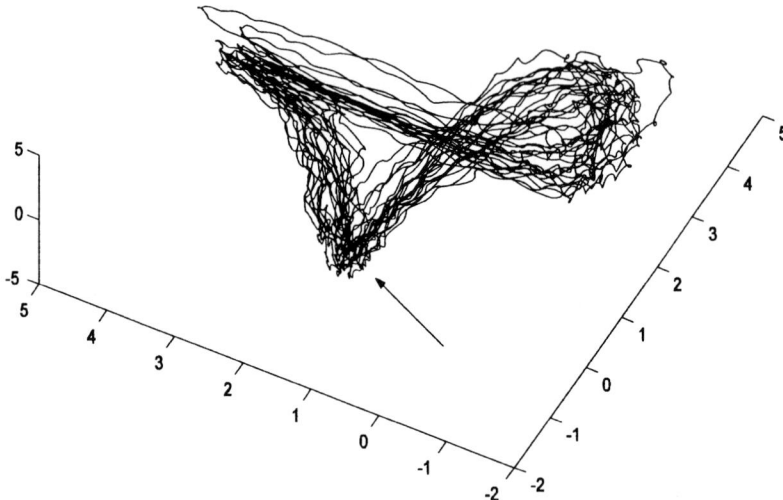

Figure 4.1. Phase plane of intratracheal pressures. Note the kink at the arrow. Several trajectories at this point are orthogonal to each other indicating their independence (singular point).

In order to minimize background noise and acquire acoustic signals as cleanly as possible, the sound recording sessions were carried out in a quiet room. Overall, the length of time of the sound recording was 30 s, and an average number of ten respiratory excursions were collected from each subject.

Normal breath sounds were recorded over the apexes by an electronically amplified stethoscope and its output was connected to a digital recorder that acquired the input signals at 44.1 kHz sampling frequency and 16 bit resolution. Simultaneously with the recording, it was possible to hear, in real-time, the acoustic signal by a headphone connected to the audio output of the digital recorder. Subsequently, each sound trace was transferred to a personal computer.

The sound files were then displayed by Fourier Transform based software that provided spectrograms (three-dimensional graphs showing the acoustic energy distribution of the signal in time and frequency domains) of the stored respiratory acoustic patterns.

This pre-analysis phase allowed for the visual identification via the spectrogram, sound traces that were affected by noise. The signals were checked for artifacts usually emanating from defective contact between acoustic sensors and neck or background noise. Contaminated segments were excluded and the best breath devoid of technical anomalies was considered suitable for further analysis.

The selected breath coming from each subject and including both inspiration and expiration phase, was then read in numeric format and band-pass filtered at 50 to 2000 Hz. The high-pass filtering at 50 Hz allowed for a reduction of muscle and heart sounds. The resulting time series data were thus used for the RQA analysis (Fig. 4.2).

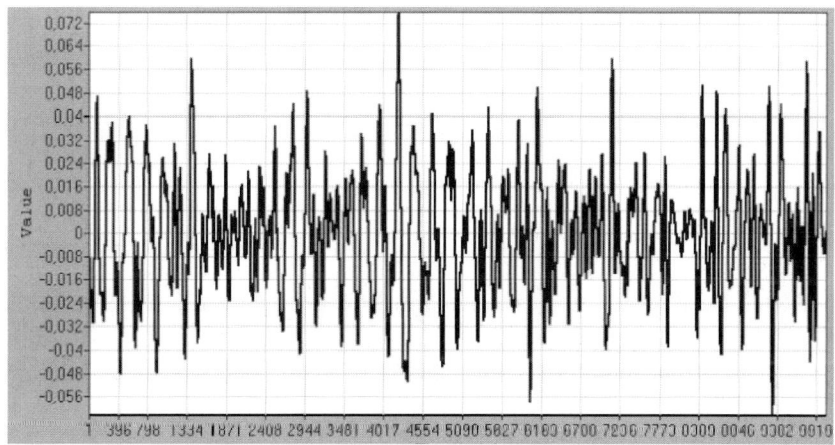

Figure 4.2. Typical lung sound recording.

As previously indicated, singularities should exhibit a divergence of the Liapunov exponent: it should tend to infinity (or in terms of RQA, maxline, a measure of system divergence). Because discrete data cannot record "infinity," a significant divergence should be recorded instead. RQA analysis demonstrated that lung sounds at the right as well as left apexes, exhibit epochs having maxline values equal to zero and thus indicating the possible presence of singularities related to nondeterministic dynamics in the respiratory system.

By inspection, it is immediately seen that singularities were detected during inspiration as well as during expiration. Figure 4.3 gives the typical behavior that was obtained for maxline (the inverse of which is related to the largest Liapunov exponent. Maxline behavior remained similar in all the seven cases.

In Fig. 4.4 another subject is presented. Note that the trajectories from left to right are broken and thus confirming that the dynamics are not continuous and not stationary but broken by singularities.

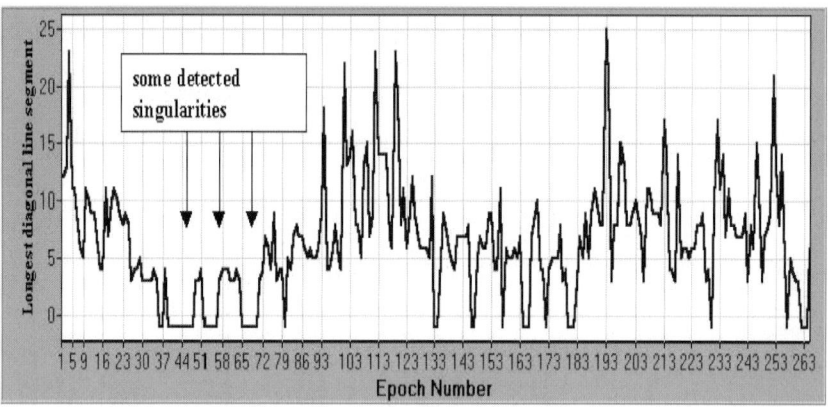

Figure 4.3. Detection of singularities in lung sounds.

In conclusion, respiratory dynamics as investigated by recording lung sounds on left and right apexes, suggest singularities, assuming a non-Lipschitz discrete behavior for respiratory dynamics. Interestingly, the sounds might conform to the cascade of opening alveoli (tiny air sacs) in the bronchial tree. This may provide the basis for further exploration into the analysis and classification of pathological sounds.

4.2.3 Heart Beat

An early and frequent subject for investigation of chaotic dynamics has been the heartbeat. Although long-standing evidence indicated that there were clear rhythmic patterns embedded in the heartbeat time series,

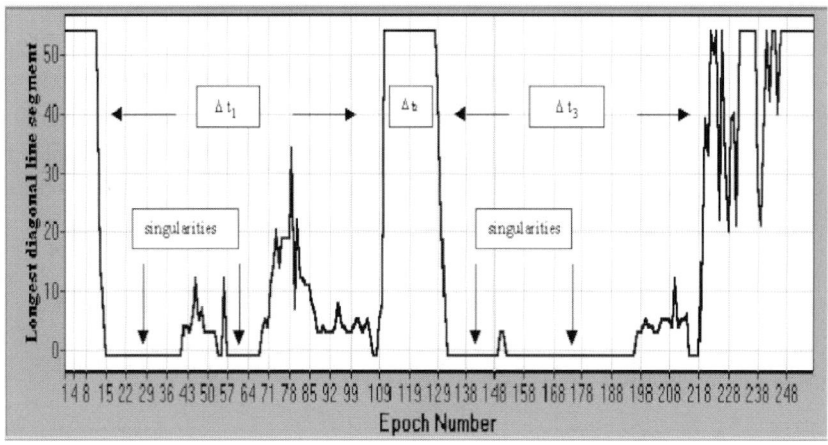

Figure 4.4. Singularities relative to inspiration/expiration (Δt)

there were also sufficient divergences from regularity to support a consideration of deterministic chaos. Even from the earliest investigations, however, there has been disagreement with this characterization, given some of the stochastic aspects of involved control processes.

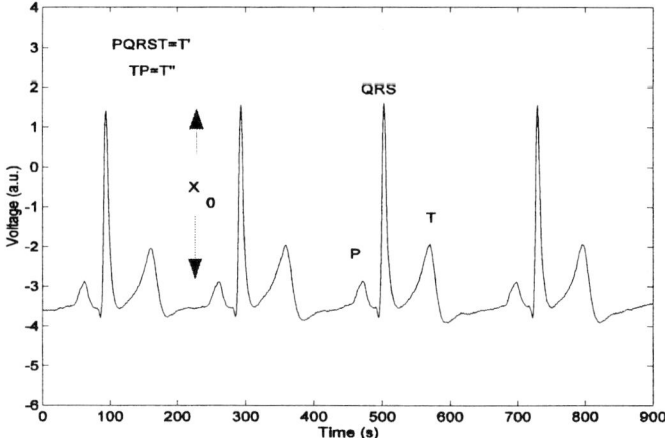

Figure 4.5. Typical ECG with "waves" labeled. T' and T'' are used for description of the model in the following figure.

Additionally, fundamental concerns regarding physiological adaptability, i.e., the ability to respond to continuous changes in perfusion requirements, have added doubts as to whether a fully deterministic system, such as chaotic dynamics can express this. It has been also noted that he heartbeat signal is composed of both deterministic, and random parts. Experiments with isolated rat hearts have suggested that heartbeats might be better described by terminal dynamics.

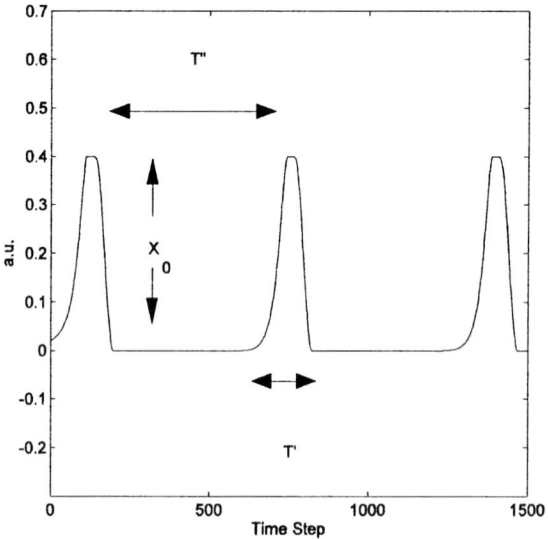

Figure 4.6. Generation of "RR" intervals from the raw ECG. T' is the normal QRS and T'' the interval.

Careful inspection of typical electrocardiograms (ECG) demonstrates that the signal is not really continuous, but rather consists of a relatively constant, deterministic portion (the PQRST complex) followed by a varying length pause, i.e., a stochastic portion (Fig. 4.5). (It should be noted that strictly speaking, there is also a pause after the P, which is beyond the scope of the present discussion). The pauses may be considered as singular points. To appreciate this, the basic ECG signal itself needs to be analyzed, and not just the R-R measurements (Fig. 4.6). The recurrence plot demonstrates that the dynamics are not continuous: a slight pause is registered at the P wave, with the continuing dynamics continuing on a slightly shifted trajectory.

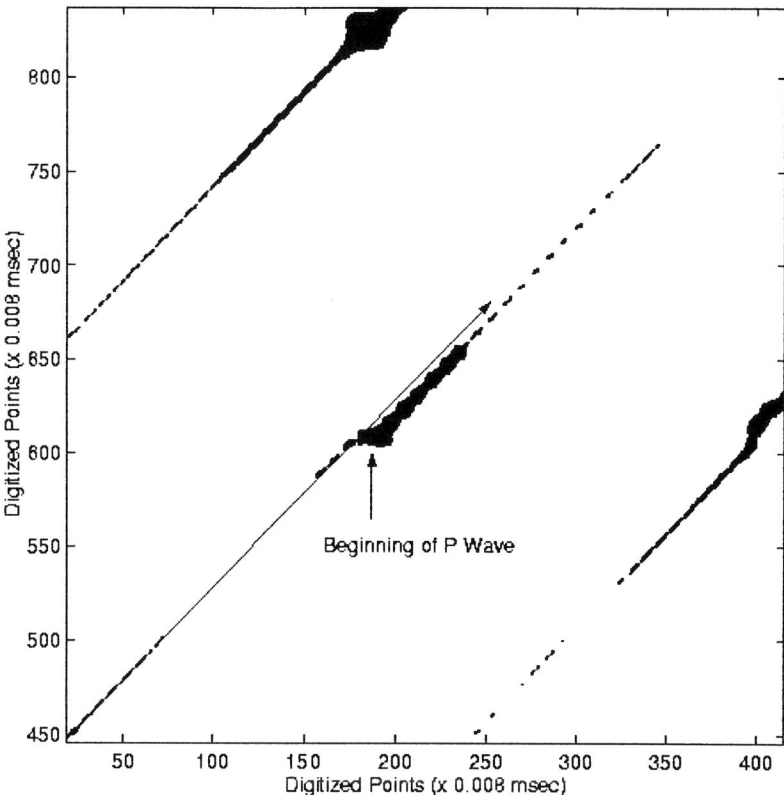

Figure 4.7. Recurrence plot of the ECG demonstrating the presence of a singularity: arrow points to a slight shift in the dynamics indicating the presence of a new trajectory

Both physiologically and numerically, a pause before the P wave has been described in the context of a resetting mechanism, or "integrate and fire" process. The full implication of such a description suggests a singular event between the T and P waves. Inspection of the ECG by an recurrence plot does demonstrate a discontinuity at this location (Figs. 4.7, 4.8). The full import of this discontinuity is beyond the scope of the present paper, but briefly, it does imply that analysis by linear and stationary nonlinear methods is not justified. It may be better modeled by non-Lipschitz differential equations whereby there is no single solution to the equation. Instead, the problem becomes a combinatorial one with a resultant "stochastic attractor." This is to say that each singularity has an associated probability distribution regulated by both intra (electrical events, stretch) and extra cardiac factors (autonomic nervous system, hormones). Thus each beat (and each subsequent beat) is conditioned by the probabilities. Based upon this phenomenon as a paradigm, a dynamical system whose solutions are stochastic processes with a prescribed joint density can be developed. By extension, sudden shifts of the dynamics as with arrhythmias may also exhibit unique densities.

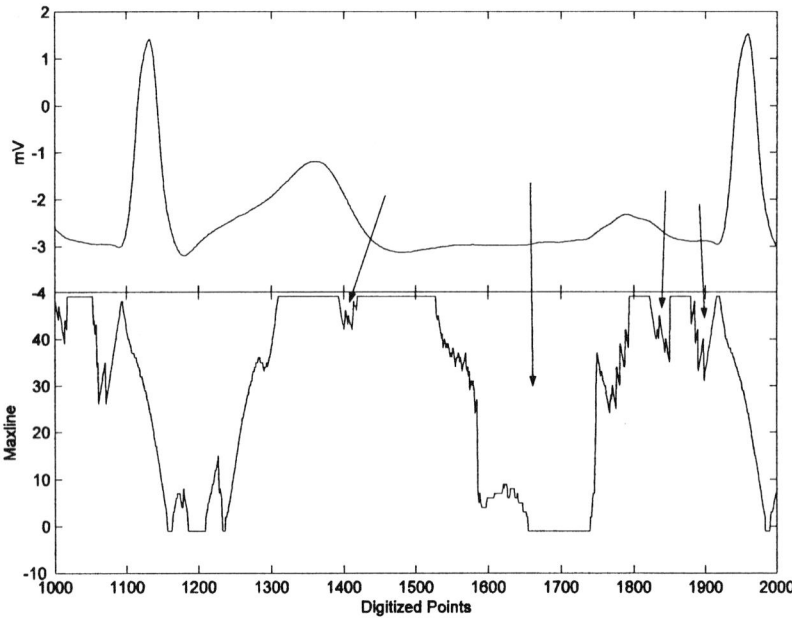

Figure 4.8. Windowed recurrence plot of the ECG for the maxline variable (see Mathematical Appendix). Arrows confirm presence of singularities. The main one occurring prior to the P wave. (The presence of other singularities is beyond the scope of the present discussion.)

A main result of this theory is that very minimal amounts of energy (and/or noise) are necessary to perform control maneuvers at the singular points. One obvious source of microscopic noise are the open-close kinetics of ion channels.

There have been very few attempts to model the effect of specific numbers of such channels. One recent attempt employed the isomorphism between open-close configurations of a set of channels, and spins on a lattice.\index{spin lattice} Of course very small amounts of noise are necessary, then, to drive the configurations on the lattice .

Simulations with small to moderate levels of noise (on the order of 10^{-4} to 10^{-8} RMS) have confirmed the absence of any significant deterioration of the dynamics (i.e., on the constant portion). This consideration is important from the viewpoint of cardiac dynamics: the heartbeat dynamics are relatively robust against noise once a beat is started, but are easily manipulated between beats. Such dynamics maintain adaptability, while preserving stability, as well as removing undue complexity.

4.3 Neural (Brain) Processes

In view of many neurobiologists, chaotic brain activity is associated with normal brain performance. However, traditional chaos, as has been emphasized, is fully deterministic, which consequently begs for an explanation for extraordinary plasticity of the brain. Nondeterministic stochastic attractors, however, significantly increase the capability of dynamical systems via more compact and effective storage and processing of information (Fig. 4.9). It is worth emphasizing a phenomenological similarity between the brain activity and singular dynamics models. Indeed, due to the singularities, the dynamical systems can be activated spontaneously driven by a global periodic rhythm. In the course of this spontaneous activity they can move from one attractor to another, change locations of attractors, create new attractors and eliminate the old ones. Another phenomenological similarity follows from the duality of the dynamical performance caused by violation of the Lipschitz condition at equilibrium points: the dynamical motion is continuous, but it is controlled by a string of numbers produced by the microdynamics: as a genetic code, the combinations of these numbers prescribe the continuous motion of the dynamical system. The multiplicity of starting points of biologically motivated neurodynamics account for evolution; whereas classical neurodynamics obliges the system to choose from a finite and pre-fixed set of behaviors.

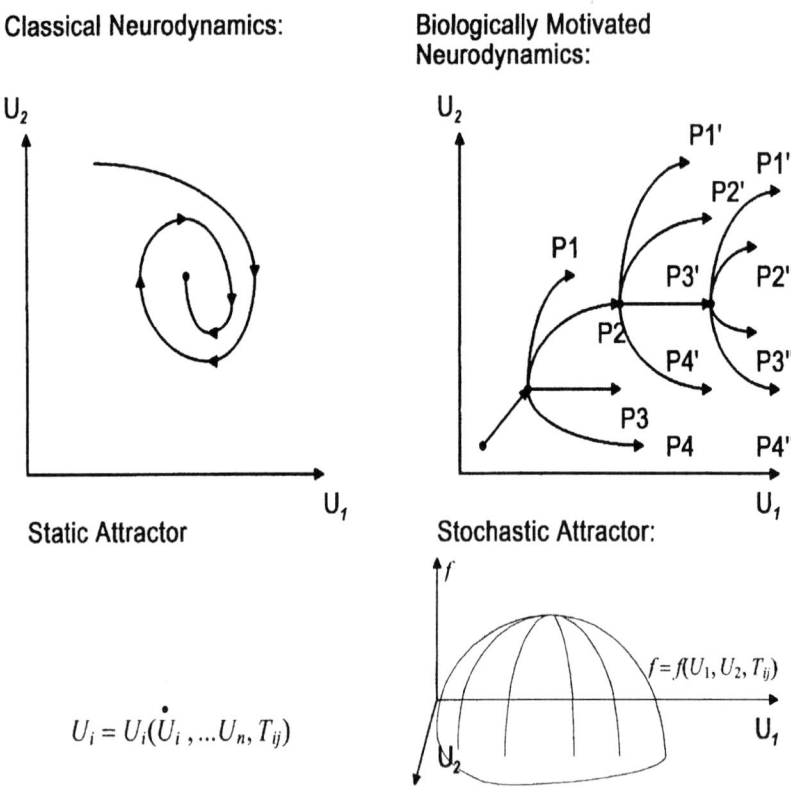

Figure 4.9. Distinction between classical and nondeterministic "neural nets": note that nondeterministic nets generate probabilities at singular points thus generating a stochastic attractor. Consequently, the problem becomes one of combinatorics of probabilities.

4.3.1 Electroencephalograms and Seizures

The neurophysiological basis of epileptogenesis has yet to be convincingly demonstrated. However, research has implicated the role of neuronal loss, changes in neuronal circuitry (excitatory mossy fiber sprouting), as well as glial cells changes (notably, impaired potassium buffering). Neuronal loss has been mostly evidenced by brain imaging. MRI studies have shown a hippocampal volume loss, which correlates with epilepsy duration. Specifically, hippocampal volume ipsilateral to the EEG focus decreased significantly as the duration of seizure increased. Similar results where found in other studies that quantified hippocampal atrophy in temporal lobe epilepsy (TLE) by MRI measurements of creatine resonance and by using the neuronal marker N-acetyl-aspartate.

In addition to the evidence of epilepsy-related neuronal injury, changes in the neuronal circuitry have been observed. It has been proposed that mossy fiber sprouting of granule cell axons within the dentate gyrus may create an excitatory feedback circuit from the granule cells back onto their own dendrites. This sprouting may then cause recurrent excitation that may lead to the generation of seizures. Recent studies indicate that progressive mossy fiber sprouting would be a response to hippocampal cell loss, and might not be just a consequence of the seizure, but rather an important factor contributing to the process of epileptogenesis.

Finally, glial cell changes have been reported in temporal lobe epilepsy. While potassium homeostasis is regulated by glia, it has been observed that when potassium influx into glial cells is impaired, an abnormal accumulation of potassium in the extra cellular space, and an increase in neuronal activity occur.

Thus, it could be suggested, based on observations of neuronal loss data, as well as changes in synaptic circuitry, and in glial cells properties, that all may affect neuronal excitability and synchronization. Clearly, these epilepsy-related brain modifications may have some consequences for the characteristics of 1) the EEG signal, and 2) seizure enhancement mechanisms.

The suggestion is that a careful analysis of EEGs may provide some evidence of the associated neurophysiological pathologies. A major difficulty, however, is that EEG signals have been classically studied with linear techniques such as power spectrum analysis. However, an important consideration is that this tool necessitates transformation of the data: the Fourier Transform (FT), for instance, represents a weighted average of transformed spectral frequencies and is based on linear systems theory. Furthermore, power spectrum analysis requires, for reliability, a relatively long, stationary signal. As a result, many attempts at time-frequency versions of the FT have been suggested with variable results. However, a drawback is that they are still based on linear systems theory. Consequently, such tools are not appropriate for EEG signal analysis, which is generated by a highly nonlinear, multidimensional system consisting of ensemble neurons. Although nonlinear techniques have been applied in an attempt to correct the drawbacks of linear methods, but once again, most of these techniques require assumptions that are not appropriate and highly debatable when biological data are analyzed: On the one hand, the finiteness of experimental time series renders the use of nonlinear chaos-based algorithms improper, and limits the ability to distinguish between signals generated by different systems. On the other hand, nonlinear measures such as Liapunov exponents and correlation dimensions, require stationarity and low dimensional determinism to give reliable results.

In this study, the alternative method, recurrence quantification analysis is used. This methodology does not transform the data, is not constrained by nonstationarity and does not make a priori assumptions regarding either the linearity/nonlinearity of the data or the system dimension. RQA simply counts the recurrence times of the plot (interval of time where 2 points are found to be recurrent). In a slightly different context, "recurrences" have also been used as supportive evidence for nonlinearity in EEGs. It also demonstrates the possibility of scaling processes as well as that of stochastic phenomena. In can do so by separating interrelated processes (such as neural firing), from autonomous processes such as noisy glial responses. In the present case, this tool was used in order to reveal scaling properties of preictal EEG segments in comparison to controls. The underlying assumption is that connections between synapses are non-Lipschitz and are easily affected by noise.

Finally, a challenging hypothesis for seizure triggering mechanism is developed based on the physics of noise interaction with nonlinear systems. Epilepsy-related brain modifications, namely, damaged neurons as well as glial cells, may create an additive abnormal activity (noise activity) which would affect the electrical activity of neuronal dynamical networks. The idea that noise could play a role in neuronal information processing has been described extensively, and specifically in the context of brain activity. This mechanism, stochastic resonance (SR), has various manifestations, but basically, the idea is that additive noise drives a nonlinear deterministic processes to amplify or alter the dynamics. Although there are many "forms" of SR, in the present context, the idea is that a change in the inherent noise characteristics of altered neurons stimulate a resonance-like phenomenon in multiply connected neurons (multiplicative process of a stochastic attractor), whereby the bandwidth of frequencies of the responding neurons is narrowed as a seizure is about to commence ("focusing"), while the noise bandwidth is at least widened.

The possibility is, then, that epileptic EEGs may be driven not only by the multiplicative process resulting from interaction between non-damaged neuronal networks (signaling), but also by some additive noise due to injured neurons and gliosis alteration. The specific mechanism by which the noise properties change are not clear, but may relate to the progressive, dynamical alterations related to, e.g., progressive mossy fiber generation, or potassium handling as suggested above, and resulting in possible baseline alteration of spiking characteristics. A critical point may be reached which triggers seizure activity (resonance), and is then subsequently shut down due to metabolic derangement, only to redevelop at another favorable time.

To study these prospects, preictal and EEG segments free of seizure activity were analyzed to study their scaling properties and to determine whether there existed different characteristics of a weak additive noise that could promote seizures: regular neurons should demonstrate "scaling" activity, whereas noisy activity should be evident but not part of the scaling process.

Since the hypothesis was that brain damage might participate in epileptogenesis, 6 patients with neuronal lesions, and associated seizure activity were chosen, out of a group of 15 consecutive admissions. Subjects were 4 women and 2 men between 25-60 years old (mean ± SD: 34 ± 11) and presented with a clear diagnosis of left temporal lobe epilepsy. Because of the EEG file storage demands of the clinical unit, only sufficient data was obtained for control and preictal segments (artifact free) for these patients. In an attempt to obtain more useful results, an additional patient was found who provided several preictal and control EEG segments.

The hypothesis that seizures could result from changes in the properties of additive noise due to neuronal and gliosis alteration was tested on preictal and control (interictal) EEG segments free of artifact. The objective was to determine if neuropathological changes were identifiable in the preictal segment as contrasted with their control.

Because the standard EEG provides 20 signals, the delay-embedding procedure is not required. Instead, the standard 20 EEG leads are first collapsed into a single time series by taking their Euclidian norm defined by:

$$\bar{x}_i = \sqrt{\sum_{i=1}^{20} |x_i|^2} \qquad (4.1)$$

where x_i represents a vector of the successive EEG value for each lead. By so doing, the need for choosing specific leads for analysis was obviated. Although this, in effect, loses spatial information, the aim was to determine if subtle EEG activity changed prior to seizures, irrespective of focus.

For the present study the actual counts of recurrences were histogrammed with each bin representing an interval of time where 2 points are found to be recurrent (recurrence times). This count was done inside a radius of 15 normalized units (prior to further processing, the signal was normalized by the unit interval) in 20-dimensional space, and the resulting log-log histogram is shown in Fig. 10. (Choice of radius is determined by the need to obtain adequate numbers of recurrences for statistical purposes, while avoiding the noise floor; and not so large so as to be sampling distant, non-local regions in multi-dimensional space. Typically, this requires approximately 1-5% recurrences in a nonstationary series.) The probability distribution that is obtained, consists of a constant part (associated with a multiplicative process: i.e., inter-correlated process of neural signaling), and of a lower part (associated with an additive noise derived from the firing characteristics of damaged neurons). The constant part is represented by the part of the data exhibiting a straight line dependence between the (log) number of recurrences (N_{Rec}) and the (log) recurrence time. This dependence can be expressed by the simple power law:

$$N_{Rec} = p^{-s}, \qquad (4.2)$$

where the exponent S is defined as the slope of the line. Indeed, the slope ends when the continuity of the scaling is interrupted and denotes the presence of a multiplicative process. The lower part corresponds to small number of recurrence times and starts after the end of the continuous part of the scaling, at a critical recurrence times (P_c). It renders the presence of additive noise. The boundary (P_c) between the multiplicative process and additive noise moves with the strength α of the additive noise in such a way that the distance from the origin (P_c) is proportional to α. Thus, a measure of changes in this scaling includes the slope, the critical time, and the bandwidths of the multiplicative and additive portions.

This scaling may possibly reveal some interesting EEG characteristics:

- a multiplicative process related to neuronal dynamics,
- the presence of additive noise, a process independent from the previous one, and
- a possible "focusing" or resonance relationship between 1 and 2.

"Focusing" is conceptually related to stochastic resonance in that noise facilitates signal detection in nonlinear systems, but stochastic resonance is typically related to external noise sources, and allows for detection of subthreshold signals with often dramatic results. Stochastic focusing, on the other hand, describes how internal fluctuations can make a gradual response mechanism work more like a threshold mechanism, with the result being less dramatic, but nonetheless meaningful.

Each time series was analyzed for its scaling properties. The slope of the multiplicative processes and the P_c were computed for preictal and control EEG segments. A slope was calculated for all contiguous points starting at the beginning, and terminated when an empty bin was encountered. The empty bin was determined to be the P_c. In this way arbitrary decisions regarding termination of scaling was avoided. A Wilcoxon Signed Ranks non-parametric test was performed to compare preictal and control groups relative to the slope and P_c. Significance was set at $p < 0.05$.

The recurrence analysis and the scaling were performed on surface epileptic EEG time series recorded with a Telefactor beehive system. The data were band pass filtered (1-70 Hz) and digitally stored with a sampling frequency of 200 Hz. Groups of EEG segments were constituted by 6 preictal times series of 6,000 points long (extracted from larger EEG segments just before the seizure onset), and by 6 EEG time series free of epileptic seizure activity (nonpreictal segments) of the same length. Preictal and control segments were not continuous and were independently selected by a neuroscientist (not involved with the analysis) with extensive experience in EEG monitoring and analysis. The only qualification was that the segments be from awake subjects with eyes open. This was done to avoid selection bias. Raw epileptic EEG data were also visually inspected for artifact.

Results show significant differences between preictal and control groups of EEG segments. As an example, Fig. 4.10 illustrates the results of the power law scaling obtained in one preictal segment (upper panel) and one control segment (bottom panel) for one patient. Thus, preictal and control EEG segments exhibit significantly different slopes ($p = 0.03$), as well as significantly different critical periods, P_c, ($p = 0.05$). It appears that preictal EEG segments have a steeper slope and a smaller P_c than control EEG segments. It is also interesting to note that the SDs of the slopes and P_cs for the control groups are smaller than those of the preictal. This may be a further indication that noise is becoming a more significant factor in the preictal circumstance; i.e., as the slopes are steeper, perforce the bandwidth of the additive noise increases.

Since the P_c delimits the boundary between multiplicative process and additive noise, these results suggest: 1) that the multiplicative process has a narrower range of high frequencies (shorter recurrence times) in preictal as compared to control EEG segments, and 2) that the additive noise has a broader range of recurrence times in preictal segments as compared to controls. The suggestion is that the broader range of noise signals a qualitative change.

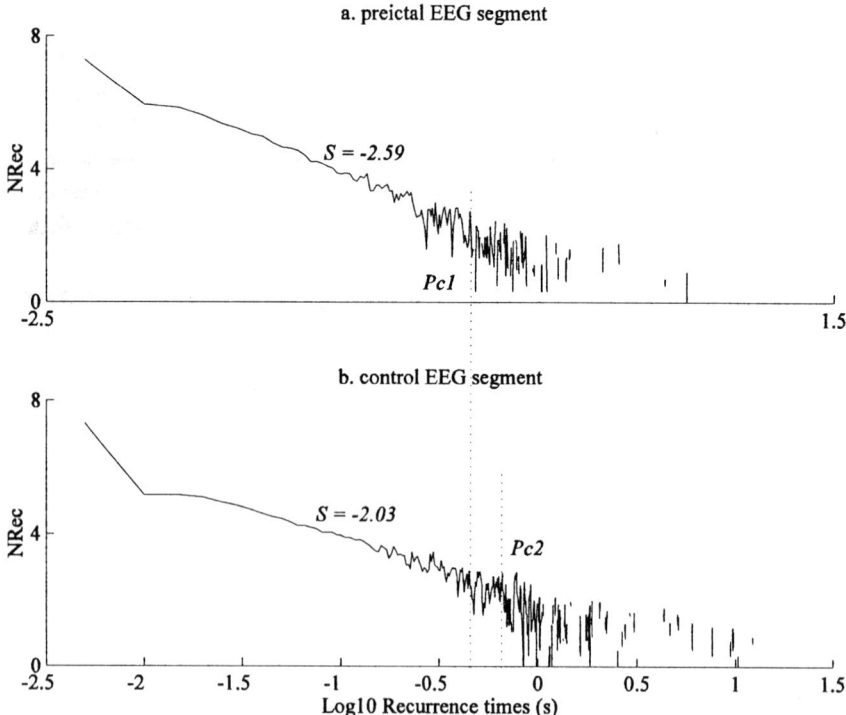

Figure 4.10. Each time series was submitted to RQA where a simple count of the recurrence times was done. An example of the resulting log-log histogram is given here. The left part of the histogram falls approximately in a straight line (S). In fact, the slope ends when the continuity of the scaling is interrupted (P_c) and denotes the presence of a multiplicative process. The lower end of the histogram starts after the end of the continuous part of the scaling, at a critical period (P_c). It renders the presence of additive noise. Example of histograms obtained in the case of (a), 30 sec preictal EEG segment and (b), 30 sec control EEG segment. Note the difference in the slope (S) and in the critical period (P_c), according to the nature of EEG activity: (a) preictal and (b) control.

Finally, the cumulative sum of recurrences at each recurrence time, 30 seconds before seizure onset were analyzed. The interval of interest consists of the first 50 recurrences times (i.e., the first 50 bins) derived from the data. The Wilcoxon Signed Rank test performed on the groups demonstrates that preictal segments have a significantly ($p = 0.03$) higher cumulative sum than control EEGs. This is to say that recurrence times tend to a more narrow frequency bandwidth with a greater amplitude; *i.e.*, a resonance-like phenomenon (Fig. 4.11).

Although these results are interesting they are limited by the fact that they are based on limited data for each patient. Consequently, the additional patient with a group of 10 preictal and 10 control EEG segments were similarly analyzed using the Kruskall-Wallis statistical test. The results were in agreement with the previous findings. Slopes are significantly steeper in preictal than in control EEG segments ($p = 0.01$). Furthermore, the additive noise appears for smaller P_c in the preictal segments than in the case of the control segments group ($p = 0.01$). Finally, cumulative sums are higher in preictal EEG segment than in control EEGs ($p = 0.04$).

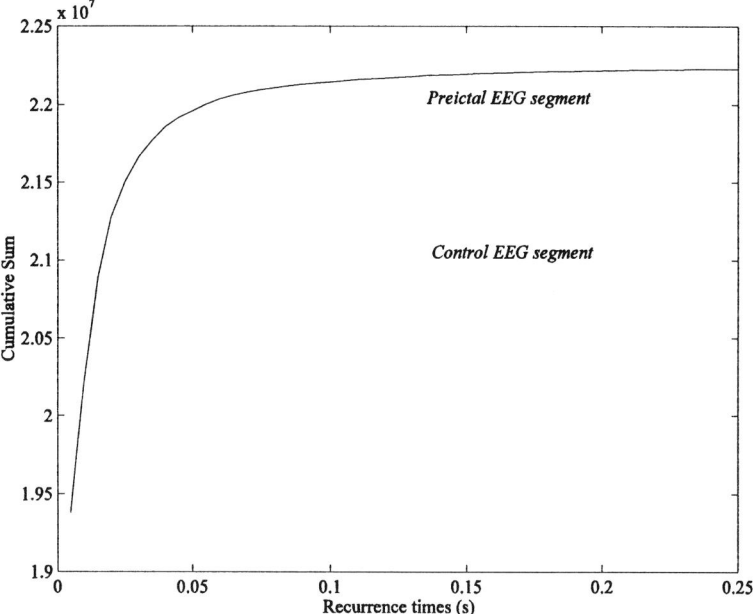

Figure 4.11. Example of "focusing" for the preictal EEG segments. For a given recurrence time, the preictal segment exhibits a larger cumulative sum than does the control.

Thus we explored the possibility that seizure may be enhanced by changes in characteristics of neuronal/glial noise. In order to test this hypothesis, we studied

EEG segment scaling properties in a group of patients presenting with a clear diagnosis of left temporal lobe epilepsy and in a patient presenting persistent seizure activity after brain damage. The modeling of this complex system was approached without strong assumptions on the one-to-one mapping between signal features and intervening biological mechanisms but with a general approach, using scaling features of recurrences which separate interrelated events from additive noise events. This work raises two main points. First, it shows that developments in the physics of noise interaction with nonlinear systems could bring new perspectives in the understanding of epilepsy. Specifically, "noise" characteristics, as generated by mechanisms such as gliosis and altered potassium buffering, may be important in the genesis of seizure activity. Second, it proposes that changes in brain cell properties might be involved in triggering the seizure onset. What initiates these changes is not yet clear. As has been shown, however, this process may be monitored by RQA and may provide warning that seizure activity is probable.

4.3.2 Terminal Neurodynamics

This section is motivated by an attempt to link two fundamental concepts of neurodynamics—irreversibility and creativity—in connection with the architecture of neural networks based upon the terminal dynamics, discussed in previously.

There is some evidence that primary mode of computation in the brain is associated with a relaxation procedure, i.e., with settling into a solution in the same way in which a dissipative dynamical system converges to an attractor. This is why many attempts were made to exploit the phenomenology of nonlinear dynamical systems for information processing as an alternative to the traditional paradigm of finite-state machines.

Over the past decade, there has been an increasing interest in dynamical aspects of artificial neural networks in the areas of global stability, adaptive pattern recognition, content-addressable memory, and cooperative-competitive decision making by neural nets. The number of publications in the area of neurodynamics is still growing exponentially. But along with this, the number of limitations of current models as well as the number of unanswered questions concerning the relevance of these models to brain performance grow too. Indeed, the biggest promise of artificial neural networks as computational tools lies in the hope that they will resemble the information processing in biological systems. However, the performance of current neural networks is still too "rigid" in comparison with even simplest biological systems. This rigidity follows from the fact that the behavior of a dynamical system is fully prescribed by initial conditions. The system never "forgets" these conditions: it carries their "burden" all the time. In contrast to this, biological systems are much more flexible: they can forget (if necessary) the past, adapting their behavior to environmental changes.

The thrust here is to discuss the substantially new type of dynamical system for modeling biological behavior introduced as nondeterministic dynamics. The approach is motivated by an attempt to remove one of the most fundamental limitations of current models of artificial neural networks—their "rigid" behavior compared to biological systems. As has been already mentioned, the mathematical roots of the rigid behavior of dynamical systems are in the uniqueness of their solutions subject to prescribed initial conditions. Such a uniqueness was very important for modeling energy transformations in mechanical, physical, and chemical systems which have inspired progress in the theory of differential equations. This is why the first concern in the theory of differential equations as well as in dynamical system theory was for the existence of a unique solution provided by so-called Lipschitz conditions. On the contrary, for information processing in brain-style fashion, the uniqueness of solutions for underlying dynamical models becomes a heavy burden which locks up their performance into a single-choice behavior.

The new architecture for artificial neural networks presented here exploits a novel paradigm in nonlinear dynamics based upon the concept of terminal attractors, terminal repellers, and terminal chaos. Due to violations of the Lipschitz conditions at certain critical points, the neural network forgets its past as soon as it approaches these points; the solution at these points branches, and the behavior of the dynamical system becomes unpredictable. Since any vanishingly small input applied at critical points causes a finite response, such an unpredictable system can be controlled by a neurodynamical device which operates by sign strings (or even noise and other appropriate signals—see above regarding seizures), and as a code, uniquely defines the system behavior by specifying the direction of the motions in the critical points. By changing the combinations of signs in the code strings, the system can reproduce any prescribed behavior to a prescribed accuracy. This is why unpredictable systems driven by sign strings are extremely flexible and are highly adaptable to environmental changes. The supersensitivity of critical points to external inputs appears to be an important tool for creating chains of coupled subsystems of different scales whose range is theoretically unlimited.

Due to existence of the critical points, the neural network becomes a weakly coupled dynamical system: its neurons (or groups of neurons) are uncoupled (and therefore, can perform parallel tasks) within the periods between the critical points, while the coordination between the independent units (i.e., the collective part of the performance) is carried out at the critical points where the neural network is fully coupled. As a part of the new architecture, weakly coupled neural networks acquire the ability to be activated not only by external inputs, but also by internal periodic rhythms. (Such a spontaneous performance resembles brain activity).

One of the most fundamental features of the new architecture is terminal chaos which incorporates an element of "arationality"/irrationality into dynamical behavior, and can be associated with the creativity of neural network performance. Terminal chaos is generated by those critical points at which the sign of the driving genetic code is not prescribed. The solution oscillates chaotically about such critical points, and its behavior qualitatively resembles "classical" chaos. However, terminal chaos is characterized by the following additional property: it has a fully predictable probabilistic structure and therefore, it can learn, and be controlled and exploited for information processing. It appears that terminal chaos significantly increases the capability of dynamical systems via a more compact and effective storage of information, and provides an ability to create a hierarchical parallelism in the neurodynamical architecture which is responsible for the tremendous degree of coordination among the individual parts of biological systems. It also implements a performance of high level cognitive processes such as formation of classes of patterns, i.e., formation of new logical forms based upon a generalization procedure.

Special attention is given to the new dynamical paradigm—a collective brain, which mimics collective purposeful activities of a set of units of intelligence. Global control of the unit activities is replaced by the probabilistic correlations between them. These correlations are learned during a long term period of performing collective tasks, and are stored in the synaptic interconnections. The model is represented by a system of ordinary differential equations with terminal attractors and repellers, and does not contain any "man-made" digital devices.

It is worth emphasizing the phenomenological similarity between brain activity and this new dynamical architecture of neural networks: due to terminal chaos, the dynamical systems can be activated spontaneously driven by a global internal periodic rhythm. In the course of such a spontaneous activity they can move from one attractor to another, change locations of attractors, create new attractors and eliminate the old ones. Another phenomenological similarity follows from the duality of the dynamical performance caused by violation of the Lipschitz condition at equilibrium points: the dynamical motion is continuous, but it is controlled by a string of numbers stored in the microdynamical device: as a code, the combinations of these numbers prescribe the continuous motion of the dynamical system. In this regard one can view the new dynamical architecture as a symbiosis of two fundamental concepts in brain modeling: analog and digital computers.

Although attention is focused on the basic properties of unpredictable neurodynamics such as irreversibility and coherent (temporal and spatial) structures in terminal chaos and the collective brain, simple computational advantages of neural nets with terminal attractors are also possible.

4.3.3 Creativity and Neurodynamics

What follows is an approach to formalization of the concept of creativity in connection with information processing in neurodynamics. One of the most important and necessary (but not sufficient) conditions for neurodynamics—to be creative, is its irreversibility. In this regard one would recall that classical dynamics is reversible within a finite time interval: trajectories of different particles never intersect, and therefore, they can be traced toward the future or toward the past. In other words, nothing can appear in the future which could not already exist in past.

This means that neural networks based upon classical dynamics are reversible, and therefore, they can perform only prescribed tasks: they cannot invent anything new. However, nature introduced to us another class of phenomena where time is one-sided. These phenomena involve nonequilibrium thermodynamics which are irreversible, and consequently, past and future play different roles. The main tool of transition to fundamentally new effects in irreversible processes is unpredictability caused by instability of equilibriums states with respect to certain types of external excitations. Hence, the unpredictability is the second necessary (but still not sufficient) condition for neurodynamics to be creative. The last condition is the utility of its new (unpredictable) performance. The simplest way to introduce this utility is to require that the new performance preserve certain global characteristics while the details are random. In other words, creative performance is not totally unpredictable: it has certain (prescribed) constraints.

The new (terminal) architecture of neurodynamics satisfies all the conditions for creativity. Indeed, it is irreversible due to violations of the Lipschitz conditions at the critical points, it is unpredictable due to terminal repellers at the critical points, and it preserves prescribed global characteristics via temporal and spatial coherence.

As has been demonstrated, the main tool for temporal coherence is terminal chaos which preserves the center of attraction as well as the probabilistic characteristics of the motion with respect to this center; at the same time, the particular values of the neuron potentials remain random until additional information (which arrives from another neuron) eliminates certain degrees of randomness converting the chaotic motion into more (or totally) deterministic motion. Hence, due to redundancy of the degrees of freedom, terminal chaos provides a tool by means of which different neurons are adjusted to each other in their parallel performance. In other words, terminal chaos represent a generalization of certain types of behaviors, or a class of patterns, while each particular member of this class can be identified as soon as additional information is available.

Different levels of complexity of temporal structures with terminal chaos are available. As has been suggested, each such a level is characterized by the number of microscales with the corresponding local times. As has been demonstrated, nonlinearities of micro-dynamics on each scale play an important role in ability of the system to move from one terminal chaotic attractor to another, as well as to eliminate one attractor and to create another one.

Spatial coherence of the neurodynamics performance is introduced via local differential interconnections represented by spatial derivatives of neuron potentials and incorporated into the microdynamics. Different levels of complexity of spatial and spatial-temporal structures of the neurodynamics performances allow for new patterns. Thus, it can be concluded that the new neural network architecture based upon terminal dynamics represents a mathematical model for irreversible neurodynamics capable of creative performance (Fig. 4.12).

4.3.4 Collective Brain

This section briefly presents and discusses physical models for simulating some aspects of neural intelligence, and in particular, the process of cognition, which can also serve as a model for collective behavior. (Is the behavior of an ant colony different from a processing brain?) The main departure from the classical approach here is in the utilization of a terminal version of classical dynamics. Based upon violations of the Lipschitz condition at equilibrium points, terminal dynamics attains two new fundamental properties: it is irreversible and nondeterministic. Special emphasis is placed on terminal neurodynamics as a particular architecture of terminal dynamics which is suitable for modeling of information flows. Terminal neurodynamics possesses a well-organized probabilistic structure which can be analytically predicted, prescribed, and controlled, and therefore, which presents a powerful tool for modeling real-life uncertainties. Two basic phenomena associated with random behavior of neurodynamical solutions are exploited. The first one is a stochastic attractor—a stable stationary stochastic process to which random solutions of closed system converge. As a model of cognition, a stochastic attractor can be viewed as a universal tool for generalization and formation of classes of patterns. The concept of a stochastic attractor can be applied to model a collective brain paradigm explaining coordination between simple units of intelligence which perform a collective task without direct exchange of information. The second fundamental phenomenon is terminal chaos which occurs in open systems. Applications of terminal chaos to information fusion as well as to explanation and modeling of coordination among neurons in biological systems can easily be obtained. It should be emphasized that all the models of terminal neurodynamics are implementable in analog devices, which means that all the cognition processes discussed are reducible to the laws of Newtonian mechanics.

4.3.5 Stochastic Attractor as a Tool for Generalization

As has been remarked, random activity in the human brain has been a subject of discussion in many publications. Interest in the problem was promoted by the discovery of strange attractors and their connection to deterministic chaos. This discovery provided a phenomenological framework for understanding electroencephalogram data in regimes of multiperiodic and random signals generated by the brain. An understanding of the role of such random states in the logical structure of human brain activity would significantly contribute not only to brain science, but also to the theory of advanced computing based upon artificial neural networks. Here, based upon properties of terminal neurodynamics, a phenomenological approach is proposed to the problem: it is suggested that a stochastic attractor incorporated in neural net models can represent a class of patterns, i.e., a collection of all those and only those patterns to which a certain concept applies. Formation of such a class is associated with higher level cognitive processes (generalization). This generalization is based upon a set of unrelated patterns represented by static attractors and associated with the domain of lower level of brain activity (perception, memory). Since a transition from a set of unrelated static attractors to the unique stochastic attractor releases many synaptic interconnections between the neurons, the formation of a class of patterns can be "motivated" by a tendency to minimize the number of such interconnections at the expense of omitting some insignificant features of individual patterns.

Consider first a deterministic dissipative nonlinear dynamical system, modeled by coupled sets of first order differential equations representing the neuron activity, and a constant matrix whose elements represent synaptic interconnections between the neurons.

The most important characteristic of neurodynamical systems is that they are dissipative, i.e., their motions, on the average, contract phase space volumes onto attractors of lower dimensionality than the original space.

So far only point attractors have been utilized in the logical structure of neural network performance: they represent stored vectors (patterns, computational objects, rules). The idea of storing patterns as point attractors of neurodynamics implies that initial configurations of neurons in some neighborhood of a memory state will be attracted to it. Hence, a point attractor (or a set of point attractors) is a paradigm for neural net performance based upon the phenomenology of nonlinear dynamical systems. This performance is associated with the domain of lower level brain activity such as perception and memory.

Thus, it can be concluded that artificial neural networks are capable of performing high level cognitive processes such as formation of classes of patterns, i.e., formation of new logical forms based upon generalization procedures. In terms of the phenomenology of nonlinear dynamics these new logical forms are represented by limit sets which are more complex than point attractors, i.e., by periodic or chaotic attractors. It is suggested that formation of classes is accompanied by elimination of a large number of extra synaptic interconnections. This means that these high level cognitive processes increase the capacity of the neural network. The procedure of formation of classes can be initiated by a tendency of the neural network to minimize the number (or the total strength) of the synaptic interconnections without a significant loss of the quality of prescribed performance; such a tendency can be incorporated into the learning dynamics which controls these interconnections.

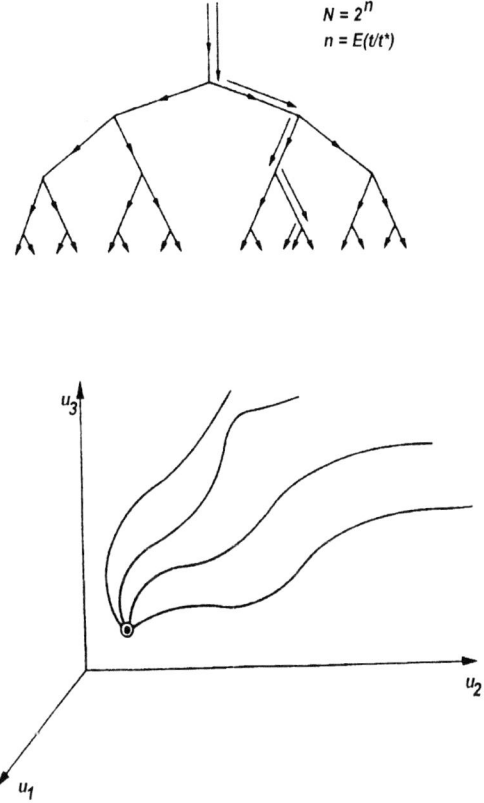

Figure 4.12. New neural patterns developed from terminal neurodynamics

In addition, the phenomenological approach presented above leads to a possible explanation of random activity of the human brain. It suggests that this activity represents the high level cognitive processes such as generalization and abstraction.

Turning to terminal neurodynamics, one can view a stochastic attractor as a more universal tool for generalization. In contradistinction to chaotic attractors of deterministic dynamics, stochastic attractors can provide any arbitrarily prescribed probability distributions by an appropriate choice of (fully deterministic!) synaptic weights. The information stored in a stochastic attractor can be measured by the entropy via the probabilistic structure of this attractor:

Random neurodynamical systems can have several, or even, infinite number of stochastic attractors. For instance, a dynamical system represents a stationary stochastic process, which attracts all the solutions with initial conditions within the area. Sequential dynamics can define a corresponding stochastic attractor with the joint density. Hence, the dynamical system is capable of discrimination between different stochastic patterns, and therefore, it performs pattern recognition on the level of classes.

4.3.6 Collective Brain Paradigm

In this section the usefulness of terminal neurodynamics, and in particular, of the new dynamical phenomenon—stochastic attractor—is demonstrated by simulating a paradigm of collective brain. The concept of the collective brain has appeared as a subject of intensive scientific discussions from theological, biological, ecological, social, and mathematical viewpoints. It can be introduced as a set of simple units of intelligence (say, neurons) which can communicate by exchange of information without explicit global control. The objectives of each unit may be partly compatible and partly contradictory, i.e., the units can cooperate or compete. The exchange of information may be at times inconsistent, often imperfect, nondeterministic, and delayed. Nevertheless, observations of working insect colonies, social systems, and scientific communities suggest that such collectives of single units appear to be very successful in achieving global objectives, as well as in learning, memorizing, generalizing and predicting, due to their flexibility, adaptability to environmental changes, and creativity.

Collective activities of a set of units of intelligence are represented by a dynamical system which imposes upon its variables different types of non-rigid constraints such as probabilistic correlations via the joint density. It is reasonable to assume that these probabilistic correlations are learned during a long-term period of performing collective tasks. Due to such correlations, each unit can predict (at least, in terms of expectations) the values of parameters characterizing the activities of its neighbors if the direct exchange of information is not available. Therefore, a set of units of intelligence possessing a "knowledge base" in the form of joint density function, is capable of performing collective purposeful tasks in the course of which the lack of information about current states of units is compensated by the predicted values characterizing these states.

This means that actually in the collective brain, global control is replaced by the probabilistic correlations between the units stored in the joint density functions.

Since classical dynamics can offer only fully deterministic constraints between the variables, we will turn to its terminal version discussed in previous sections. Based upon the stochastic attractor phenomenon as a paradigm, we posit a dynamical system whose solutions are stochastic processes with a prescribed joint density. Such a dynamical system introduces more sophisticated relationships between its variables which resemble those in biological or social systems, and it can represent a mathematical model for the knowledge base of the collective brain.

4.3.7 Model of Collective Brain

Let us first turn to an example and consider a basketball team. One of the most significant properties necessary for success in games is the ability of each player to predict actions of his partners even if they are out of his visual field. Obviously, such an ability should be developed in the course of training. Hence, the collective brain can be introduced as a set of simple units of intelligence which achieve the objective of the team without explicit global control; actions of the units are coordinated by ability to predict the values of parameters characterizing the activities of their partners based upon the knowledge acquired and stored during long-time experience of performing similar collective tasks.

A mathematical formulation of the collective brain for a set of two units are:

$$\dot{x}_1 = \gamma_1 \sin^k \left| \sqrt{\omega} \sin(x_1 + x_2) \right| \sin \omega t , \qquad (4.3)$$

$$\dot{x}_2 = \gamma_2 \sin^k \left| \sqrt{\omega} \sin(x_1 + x_2) \right| \sin \omega t . \qquad (4.4)$$

This system has a random solution which eventually approaches a stationary stochastic process with the joint probability density

$$f(x_1, x_2) = 0.5 |\cos(x_1 + x_2) \cos(x_1 - x_2)| \qquad (4.5).$$

As follows from this solution, one can find the probability density characterizing the behavior of one unit (say x_1) given the behavior of another, i.e., x_2.

Let us assume now that the unit x_1 does not have information about the behavior of the unit x_2. Then the unit x_1 will turn to the solution for the probability density which is supposed to be stored in its memory. From this solution, the conditional expectation of x_2 given x_1 can be found.

It should be noted that the general case represents a collective brain derived from the original system which are self-contained: they are formally independent since the actual contribution of the other unit is replaced by the "memory" of its typical performance during previous (similar) collective tasks. This memory is extracted from the joint probability density in the form of the conditional expectations.

It should be stressed that the main advantage of the collective brain is in its universality: it can perform purposeful activity without global control, i.e. with only partial exchange of information between its units. This is why the collective brain can model many collective tasks which occur in real life. Obviously the new tasks are supposed to belong to the same class for which the units were trained. In other words, too many novelties in a new task may reduce the effectiveness of the collective brain.

4.3.8 Terminal Comments

This brief recapitulation of the theory of terminal dynamics for neural processes has presented physical models for simulating some aspects of neural intelligence, and in particular, the process of cognition. The main departure from classical approach here is in utilization of terminal version of classical dynamics. Based upon violations of the Lipschitz condition at equilibrium points, terminal dynamics attains two new fundamental properties: it is irreversible and nondeterministic. Special attention was focused on terminal neurodynamics as a particular architecture of terminal dynamics which is suitable for modeling of information flows. Terminal neurodynamics possesses a well-organized probabilistic structure which can be analytically predicted, prescribed, and controlled, and therefore, which presents a powerful tool for modeling real-life uncertainties. Two basic phenomena associated with random behavior of neurodynamical solutions are exploited. The first one is a stochastic attractor—a stable stationary stochastic process to which random solutions of closed system converge. As a model of cognition, a stochastic attractor can be viewed as a universal tool for generalization and formation of classes of patterns. The concept of stochastic attractor was applied to model a collective brain paradigm explaining coordination between simple units of intelligence which perform a collective task without direct exchange of information. The second fundamental phenomenon discussed was terminal chaos which occurs in open systems. Applications of terminal chaos to information fusion as well as to explanation and modeling of coordination among neurons in biological systems were discussed. It should be emphasized that all the models of terminal neurodynamics are implementable in analog devices, which means that all the cognition processes discussed are reducible to the laws of Newtonian mechanics.

One of the immediate practical applications of terminal neurodynamics is the model of the collective brain which mimics collective purposeful activities of a set of units of intelligence without global control. Actually, global control is replaced by the probabilistic correlations between the units. These correlations are learned during a long term period of performing collective tasks, and they are stored in the joint density function. Due to such correlations, each unit can predict the values of parameters characterizing the activities of its neighbors without direct exchange of information. The model of the collective brain describes more sophisticated relationships in dynamics which are typical for biological and social rather than physical systems. In particular, this model can be effective in distributed control and decision analysis. With respect to the last application, the model of collective brain offers a compromise between two polar interpretations of the concept of probability as to whether it is a state of mind or a state of things. Indeed, in the course of training, the units of intelligence in the collective brain learn the objective relationships between them which are represented by physical probability distribution; this distribution is stored in the synaptic interconnections and becomes a property of each unit thereby representing personal or subjective probability.

4.4 Arm Motion

A human arm was placed in a device such that the elbow was fixed and the forearm was supported by the device and allowed to move freely in the transverse plane. The subject was instructed to move the arm slowly back and forth at a comfortable pace. The movement was digitized at 10 kHz with 2048 bits of resolution in arbitrary units. The resultant activity produced sine wave time series (Fig. 4.13). For a second realization the signal was filtered using a bi-directional low-pass, sixth order Butterworth filter at 2 Hz. Another version of

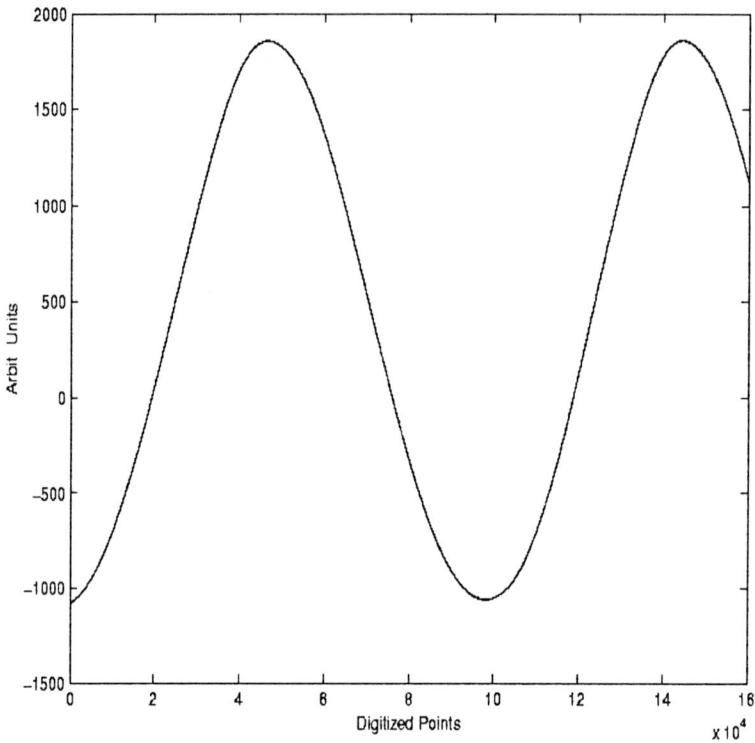

Figure 4.13. Recorded arm motion

the signal was realized by down sampling by a factor of eight.

The signals were subjected to a sliding window form of recurrence quantification analysis for determination of the maxline variable, the inverse of which approximates the largest positive Liapunov exponent.

According to previous analyses, singularities should exhibit a divergence of the Liapunov exponent. When compared to deterministic chaotic systems where an initial volume is stretched and folded, spreading across an attractor in smooth fashion, the dynamics characterized by encountering a singularity scatter points randomly throughout a region of phase space.

In contrast to a simple sine wave, the arm motion clearly demonstrates a divergence of the maxline (Fig. 4.14). This is to be expected given that at the end-point of each motion, a slight "pause" occurs, although the time series does not clearly indicate this. The filtered form of the arm motion, however, severely attenuates the divergence, and the decimated form demonstrates no divergence. This is contrasted with the results of the second derivative: noisy and no evidence for a divergence.

Clearly, maxline divergence occurs when singularities are encountered. However, the presence of a divergence may not be sufficient to posit the existence of a singularity. Such divergences can occur in regular or transient dynamics, simply indicating a local change. Assumptions regarding singularities need further support in the form of inspection of the time series, as well as a reasonable justification for its existence. Indeed, if there is reason to suspect such dynamics, unnecessary filtering should be avoided, and A/D sampling should proceed at a rate as high as practicable. Where possible, detection of orthogonal trajectories of phase plane portraits, and examination of dynamical behavior in the presence of noise can add further evidence to sustain a conjecture of singularity.

Given the difficulty of determining singularities, it is natural to question the utility of knowing of their existence. Consideration of the arm motion is instructive. Although the basic deterministic pattern is known (i.e., the sine-like waveform), technically, the pattern is unpredictable due to the variable "pause" of the singularity. In other words, the uncertainty about the precise instant at which the arm re-starts its motion turns a theoretically predictable phenomenon into a practically unpredictable one. Indeed the long-term future motion can only be described in a statistical sense. This is due to the "explosion" of information at the singular point. Additionally, control of such systems, although similar to that of chaotic systems, is somewhat different: exquisite control is obtained near the singularity where infinitesimal perturbation can effect "motion."

Perhaps the most important implication for such systems is in terms of the energetics of control. The greatest efficiency is obtained near singular points. At the same time, noise and other modulations can work profitably at the singularities (in the context of biological signals) to maintain adaptability, while trajectories outside the singularities are essentially robust against noise perturbation to maintain stability. It would seem such systems with interconnected singular points provide the opportunity for flexible control at minimal effort.

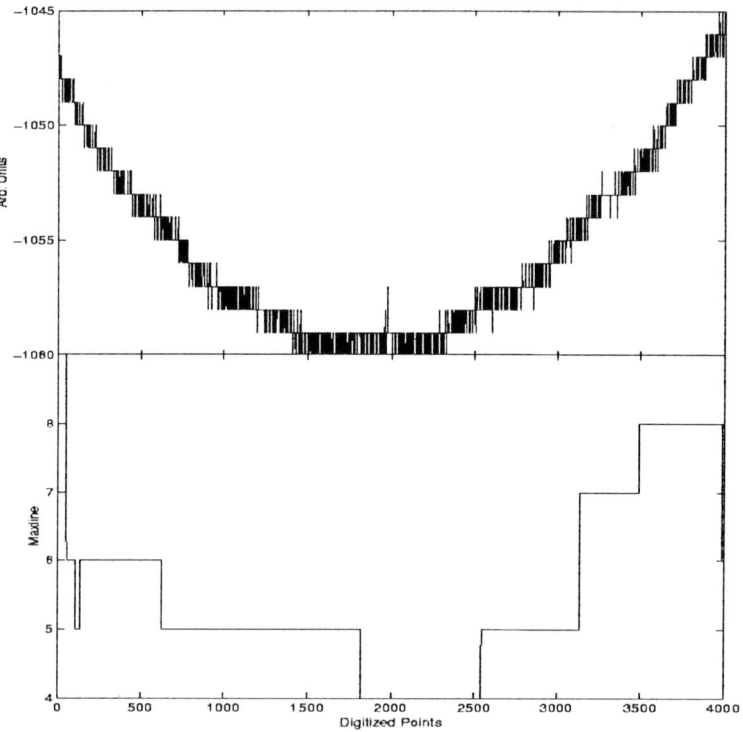

Figure 4.14. Detail of arm movement at an extremum (above), with a clear divergence (below) of the approximate Liapunov exponent at the extremum (singularity

4.5 Protein Folding

Although heretofore the theory of singularities has been presented in terms of time series, in fact, any ordered series may be considered. Relative to proteins, their sequence of amino acids "implies" their dynamics in an aqueous environment.

From a chemical viewpoint, proteins are linear heteropolymers that, unlike most synthetic polymers, which are condensed from one or a few monomer units, are formed by basically non-periodic juxtaposition of 20 different monomers. (We do note that there are some periodicities associated with alpha/beta structures, but these are of the most elementary kind, which frequently do not require sophisticated algorithms for their discovery.) A further distinction is found in their structural organization: while polymers are generally very large extended molecules forming a matrix, the majority of proteins fold as relatively small self-contained structures. Each protein found in nature has a specific three-dimensional structure in vivo and this structure is determined by the sequence of monomers. This makes the linear arrangement of amino acids constituting a protein an efficient recipe for the solution of a chemico-physical problem that basically is the folding to a given unique 3D structure in water solution. (For the sake of simplicity, here we refer only to globular proteins, the transmembrane proteins living in two different lipid and aqueous environments, in some sense solve a different problem). Obviously proteins must "solve" other problems too, like the catalysis of a particular reaction or the coordination of different ligands and effectors, but these problems are typically solved by specialized portions of the protein structures (e.g. active site). What the sequence at large must attain in order to be a real protein is basically to be water-soluble, having a well-defined (if not necessarily static) 3D structure allowing for motions in solution (proteins do their work in a dynamical way by coordinated motions of their scaffolds) while at the same time maintaining their global shape. This is not an easy task and only a relative minority of linear amino acid arrangements are effective solutions to this problem. This is equivalent to saying that the "code" linking a sequence to its particular structure is not apparent in simple periodicities of amino acid use, which, on the contrary, is evident in fibrous proteins forming large, insoluble matrices such as collagen, and many other proteins which can form aggregates instead. Nevertheless we know, from the possibility of sequentially denaturing and re-folding a given protein, that the three-dimensional structure of a protein is in some (still obscure) way encoded in its amino acid order. Indeed, this question has become more intriguing given the recent demonstration and suggestion that any protein can fold into an aggregate as opposed to its functional form. Thus, the most basic problem in the sequence/structure puzzle is: "What particular linear arrangement of amino acids makes a real protein?"

4.5.1 *Two General Contemporary Schemata*

There is a huge literature dealing with theoretical models trying to implement ab-initio modeling of the sequence/structure puzzle. The great majority of the theoretical models adopt a statistical physics perspective based on proteins considered as lattices (or even off-lattice); i.e., squared grids in which each particular residue is considered as interacting with the same number of neighbors surrounding it. Minimal models have been devised in hopes of capturing the important aspects of proteins by separating amino acids into hydrophobic and polar categories, with the emphasis that such simple models can capture many important features of proteins, including cooperativity, folding kinetics, and designability of protein structures.

Other scientists adopt an alternative view to the sequence/structure puzzle: instead of looking for "universals;" i.e., for general laws linking sequence and structure across all protein families, they apply a purely local statistical approach. A protein of which only the primary structure is known is compared, in terms of relative sequence alignment, with a huge number of already known protein sequences of which the three-dimensional structure is resolved. The scoring of a significant superposition between the query sequence and an already resolved protein allows for structural and consequently functional, (but this point requires some further specifications given the non perfect one-to-one mapping between structure and function) inferences about the investigated protein. The refinement of new sequence alignment techniques is one of the basic pillars of "bioinformatics" science. The purely statistical character of this approach prevents direct physico-chemical hints about protein folding mechanisms to be derived along this line.

4.5.2 *A Different View*

The approach we advocate is half-way between theoretical-intensive statistical physics models and purely empirical bioinformatics approaches to sequence/structure relations. These middle-way approaches are adopted by a small minority of protein scientists, but may be poised to take advantage of both physical and informatics views. The basic feature of these approaches is the adoption of a "medicinal chemist" point of view by downsizing the sequence/structure problem to the level of homogeneous series of proteins, analogously to organic chemistry congeneric series. In other words, instead of looking for the general reconstruction of fold from sequence-based features, the goal is to try and model structural and physiological variations among different elements of the series by means of corresponding variations of sequence features, according to a strategy similar to the modulation of pharmacological potency obtained through relatively small variations imposed on a lead drug. This "middle way" is achieved by mixing together elements coming from the two above mentioned extremes: from the theoretical side comes the consideration of protein sequence as a unitary system embedded into a global force field based on hydrophobicity; and from the statistical side the local approach and the use of statistical approaches not imposing any peculiar distributional constraints on the data. Moreover, the relative success of minimalistic models such as lattice models, highlights the possibility of capturing the essentials of the protein sequence/structure puzzle by approximate models. Our unique addition to this perspective has been the use of nonlinear signal analysis techniques to elucidate patterns/singularities of hydrophobicity.

4.5.3 *Proteins from a Signal Analysis Perspective*

When coded as monodimensional arrays of hydrophobicity values corresponding to the amino acid sequence translated in terms of one of the many possible hydrophobicity scales, the primary structure of a protein can be considered as a numerical discrete series equivalent to a time series with amino acid order playing the role of subsequent time intervals. And thus, just as traditional time series can exhibit unique "signatures," say, using power spectra, so can hydrophobicity series. Thus, from a purely theoretical point of view, any one of the myriads of techniques routinely used for signal analysis in diverse physical sciences could be profitably applied to protein hydrophobicity sequences. From a practical viewpoint, the fact that protein sequences are short with respect to the signals routinely analyzed in other fields (even if there is no reasonable physical limit to the length of a polypeptide chain, there are few naturally occurring proteins with more than 1000 residues) and in some cases extremely short (rubredoxins, a class of proteins which are formed by only 50 amino acids, and even shorter protein sequences do exist) drastically limits the range of signal analysis techniques we can use in the analysis of protein primary structures. Moreover protein sequences are basically non-stationary signals displaying different statistical and correlation features along the chain, thus limiting the applicability of more classic signal analysis techniques like Fourier analysis.

Ideal methods for approaching signal analysis of protein sequences should be able to deal with nonlinearity, independent of any stationary assumptions and able to deal with very short series. The method that we developed, to satisfy these constraints, we term recurrence quantification analysis. It approaches the series from a purely correlative point of view, without placing on the studied series any pre-formed distributional and/or physical assumptions.

4.5.4 Singularities of Protein Hydrophobicity

Simple models of protein folding on a lattice have been useful for the understanding of basic organizing principles of protein folding dynamics. One important idea gleaned from these studies has been the conjecture that polymers may have multiple ground states, and thus may fold into different structures.

Some investigators have extended this idea in combination with kinetic partitioning to suggest a possible phenomenology for the conformational flips of prions. They designed sequences of lattice model proteins which exhibited two different conformations of equal energy corresponding to the global energy minimum. Folding simulations demonstrated that one of these ground states was much more accessible than the other. A critical factor in determining the accessibility was the number and strength of local contacts in the ground state conformation. Although it is recognized that this may not be the only factor involved in such a phenomenology, it does provide some basic understanding of the process. To explore this possibility, as well as the feasibility of deriving an empirical, hydrophobicity based phenomenology, RQA of hydrophobicity values was applied along the sequence of the two given

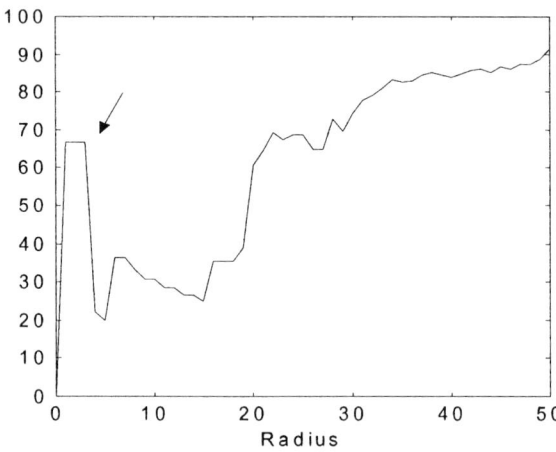

Figure 4.15. Arrow points to a singularity of the hydrophobicity profile (narrow shelf of relatively high determinism).

model 36-mers. The results were compared to the recombinant prion protein (PrP) of the Syrian hamster, shPrP(90-231) (1B10), which corresponds to the infectious fragment of the scrapie isoform.

In the present analysis, the determinism (DET, percentage of recurrent points forming line segments) was calculated for a radius from 1 to 100% (the maximum; distances being rescaled on the unit interval) with an embedding of 3 to simulate a chemical environment in which each residue "views" adjacent residues in simulated three dimensions. (It is emphasized that these dimensional perspectives should not be confused with real coordinates. The dimensions are a result of the mathematical "embedding" procedure. As a control, the sequences were randomized 25 times, with a resultant loss of the divergence. This implies that the particular hydrophobicity ordering along the sequence entails important information about the protein folding process.

A similar RQA was also performed for the Syrian hamster PrP sequence of hydrophobicity values. In order to see the change of determinism along the entire shPrP sequence, a form of RQA was performed similar to the windowing procedure common in spectral analysis. Windows of 36 residue values (to effect a 36-mer) were stepped through the sequence, overlapping one residue at a time.

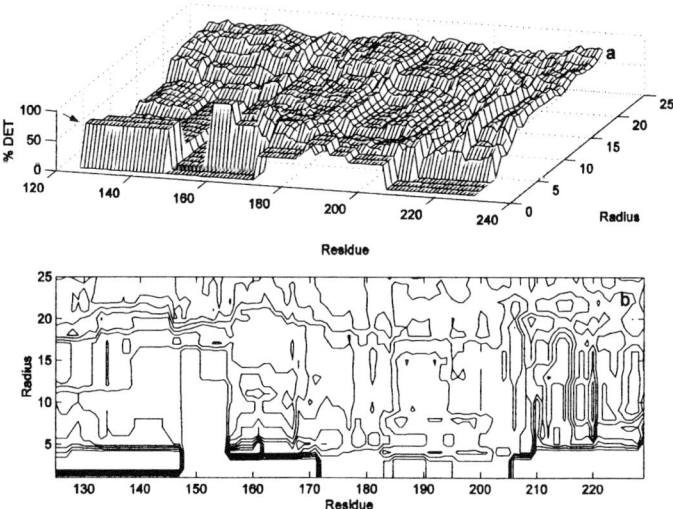

Figure 4.16. Arrow point to a singularity in sliding window RQA analysis of p53 (a). It is characterized by a "shelf" of high determinism. In (b) the corresponding contour plot is presented.

For the two model 36-mers, the results demonstrate a shelf-like divergence between relatively linear constant DET values in the low radius region, which quickly drop off to become exponentially increasing values (Fig. 4.15). A similar shelf is constant from residues 127-149 (Fig. 4.16). The implications would seem straightforward: contrary to the impressions of hydrophobicity plots, which suggest no remarkable features, RQA demonstrates that there is a definite, pronounced structured area (high values of DET) if one considers their apparent closeness (low radius) in embedding space. This structuring should be understood in terms of a repetitive hydrophobic/hydrophilic pattern, and not simply as a region of uniform hydrophobicity values. What is more striking is the narrowness of the shelf and concomitant drop off. This would imply that local contacts predominate.

It is important to emphasize that the local character of the contacts means that residues *close in space*, (in the usual 3D Euclidean space), constitute a nucleation center driving the subsequent folding of the entire protein. Here we add another dimension to the closeness in the Euclidean geometrical space: the closeness in the hydrophobic distribution space, which is the one investigated by RQA. The nucleation center is identified by RQA as a localized (Euclidean space) singularity in the hydrophobic ordering of residues (chemico-physical space) thus providing a mechanistic interpretation of the observed folding behavior.

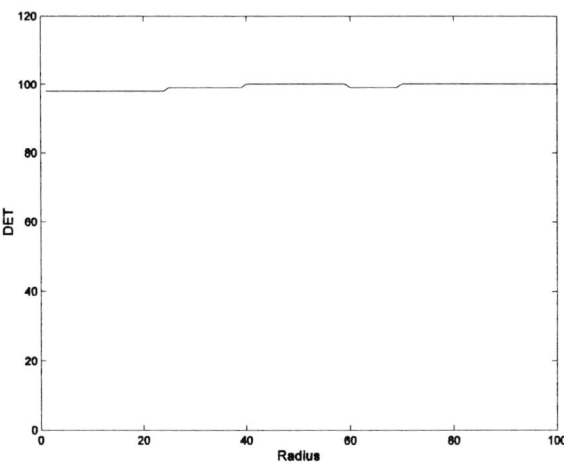

Figure 4.17. Singularity of a different kind: constant high shelf of determinism of hydrophobicity.

As such, this area may be termed singular, insofar as after the fall-off, the DET values increase slowly with no unique profile. In one sense, this singularity is unstable. If there are perturbations which can sufficiently destabilize (e.g., ΔpH, Δtemperature, or mutations) this arrangement, the observed ordered hydrophobicity can be easily destroyed. A different folding could then develop with "access" to DET patterns beyond the shelf. Presumably, this would increase the time to reach such a different state.

The RQA results for the recombinant Syrian hamster PrP revealed a divergence similar to one found in the model proteins, and is immediately adjacent to the flexible region of residues 29-124. Some investigators have remarked that residues 90-231 are neither particularly hydrophobic nor hydrophilic.

We have explored other such models, and confirm this singularity. What is more interesting, some researchers have suggested that there may be multiple folding funnels (as opposed to local minima) for proteins. In our RQA analysis of this model, we have found a slightly different form of the singularity: the singularity starts at a high value and remains high at increasing radius (Fig. 4.17). We have seen such singularities in some proteins, and they may represent a unique signature for multiple funnels. Their significance within the broader protein profile remains to be elucidated.

An almost ideal model system to confirm these findings is represented by the acylphosphatase (AcP protein, whose aggregation propensity was carefully analyzed previously. These authors demonstrated the presence of mutational "aggregation zones" along AcP the sequence in the 16-31 and 87-98 residues range: only mutations intervening in these portions of the sequence are capable of influencing the aggregation behavior of the protein. When looking at the hydropathy profile of AcP no unique feature of the curve characterizes this site. On the contrary, when submitted to the windowed version of RQA, with delay = 1, emb = 3, epoch = 28, shift = 1, scaling = maxdist, radius = 30 a unique determinism peak is evident at one of the "aggregation sensitive" portion of the sequence (Fig. 4.18).

This result was augmented by the analysis of the mutations.: using a strategy previously employed the change in recurrence values were computed for all given mutations, and found to be significantly different between the aggregation zones and folding zones ($p = .005$; Kruskal-Wallis). When submitted to a logistic discriminant function, the results were also significant ($p = .04$).

The ability of variation in recurrence properties of the sequences (or better of portion of sequences) to predict the aggregation behavior of particular protein systems was confirmed by the analysis of another specific case: the fluorescent proteins (FP) of *Anthozoa*. These proteins are used as *in vivo* markers of gene expression, protein interactions and trafficking, as well as molecular sensors. One disadvantage of the FP based techniques is linked to the aggregation properties of these proteins that may impede all possible applications due to considerable cellular toxicity. Investigators employed site-directed mutagenesis of residues located near the N-termini of FPs to resolve the problem of aggregation.

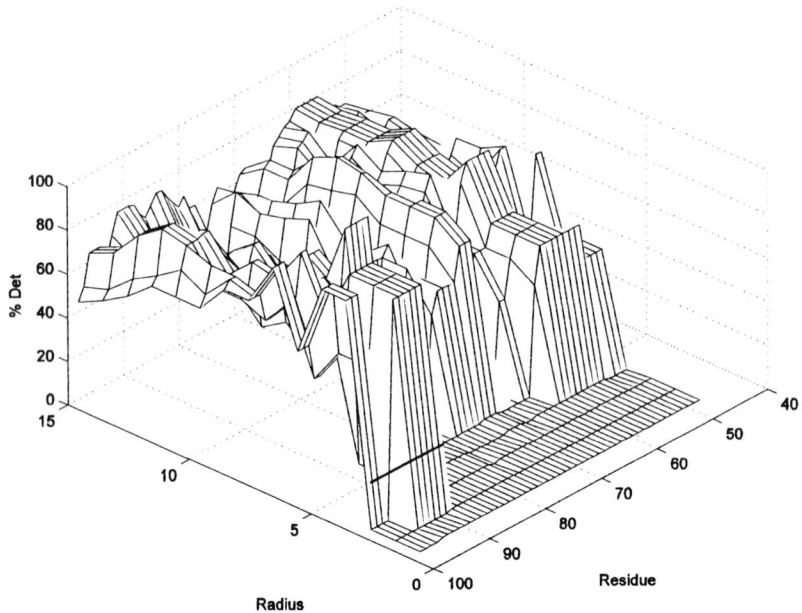

Figure 4.18. Ensemble DET for AcP. The "aggregation zones" are identified by the solid bar. Note that they are characterized by singularities. Here, residues 87-98 are highlighted.

For each of 6 parental FPs they were able to produce a non-aggregating mutant by changing few aminoacids. Thus for each FP there is an aggregating / non-aggregating pair. RQA was applied at these sequences obtaining values for the

REC of the different pairs. An important observation to note that the aggregating-to-nonaggregating shift involves a change in recurrence of the same sign ($p < 0.05$ at Paired t-test) so giving an independent confirmation to AcP data about the importance of this parameter for aggregation prediction.

What are the consequences of the presence of a singularity in the hydrophobicity pattern of consecutive residues for the dynamics of the protein structure ? As a matter of fact what we need to know are the consequences in terms of relative flexibility (governing the kind of intermolecular link the protein will form) of the structure when in presence of a given local hydrophobicity distribution pattern of a set of consecutive residues. A general statistical guess we can safely assume is that extremely repetitive (low complexity) portion of sequences are very likely to be natively disordered and thus, at least in principle, very flexible and prone to interact with other structures. This guess is sustained by a relevant literature and, was confirmed in terms of RQA parameters: a random sample of 1141 proteins from SWISS PROT was ordered by means of a principal component derived consensus score of percent recurrence of sequences according to different aminoacidic codes (hydrophobicity, polarity, volume, molar refractivity, molecular weight, literal code) : the 10 most recurrent (in terms of REC) proteins of the set were estimated to have an average 75% of native structural disorder, i.e. they are predicted to not assume a fixed tertiary structure in solution. This datum must be compared with an average 31% of predicted disorder for the 10 proteins at the other extreme of the scale (t test $p < 0.00001$). This result implies that low complexity (extremely high periodicity of subsequent aminoacid residues) is a sufficient (albeit not necessary) condition of high flexibility of the structure.

This general datum is mirrored by consideration of the highly deterministic, natively disordered portion of P53, i.e., the so called trans-activation domain, which is the portion of the protein sequence influencing (by interaction with) the behavior of other P53 molecules as well as the link with both other proteins and DNA. This piece of evidence connects the protein aggregation problem to the protein-protein interaction problem, that was recently demonstrated to be predictable in terms of recurrence parameters.

A more general treatment of singularities with a links to the protein aggregation case can be made: We have argued that there exist many physical and biological "motions" which may be better modeled by non-Lipschitz (non-smooth) differential equations whereby there is no single solution to the equation. In plain terms, a non-smooth (and thus non-Lipschitz) approach implies the existence of discrete and identifiable "turning points" in which the observed phenomenon instead of approaching in an infinite time to a limit (as in classical differential equations), on the contrary, is abruptly stopped. Examples of these kinds of dynamics are the "rip" of a flag or the discharge of a seismic wave. In all these systems we can identify "singular" points in which the dynamics abruptly "forget the past" and re-start with a completely stochastic choice of direction. In this situation the relative probability of moving one way or another becomes a combinatorial one with a resultant "nondeterministic" chaos, and "stochastic attractor." This is to say that there exist unstable singularities with an associated probability distribution regulated by factors unique to the given system. In the protein case, narrow patches of high recurrence and/or determinism at low radius are the basis of this singularity. Based upon this phenomenon as a paradigm, a dynamical system whose solutions are stochastic processes with a prescribed joint probability density can be developed. Although in the present case the discussion centers around a topological feature, without loss of generalization, it can be argued that the energy functions of molecular dynamics behave in a similar way. This is to say that if the process is described as a minimization of the energy function through potential minima, the boundaries can be described as unstable singularities as well. In the long run, the associated probabilities may govern whether a given protein will go to its native fold or an aggregation, and may be simply described by a binomial distribution. The probabilities themselves are governed by the boundary conditions; i.e., pH, temperature, etc. (Fig. 4.19).

Protein aggregation seems to be governed by the same laws of protein folding and protein-protein transient interaction. At an "extreme" of this continuum we find the polymerization of protein monomers to generate highly structured materials. At the basis of all these phenomena is the possibility to different portions of aminoacid sequence to mutually recognize and interact. The consideration of local periodicity patterns of aminoacid residues along the protein sequence is of paramount importance to judge the aggregation propensity of the relative structure, this was demonstrated at different levels of definition going from the extremely periodic hydrophobicity patterns of polymerizing proteins to the discovery of punctual local peaks of periodicity in globular proteins that have a given probability to give rise to aggregation. More in general the shift between correct folding and misfolding leading to aggregation can be considered as a basic stochastic process. This stochasticity is probably at the basis of the pathogenesis of all the so called misfolding diseases influencing macroscopic features of these like the slow onset and age dependence. And increasingly there is a growing awareness that given the correct boundary conditions any protein will fold or aggregate. Given the importance placed on oxidative stress for physiological degeneration, it is easy to contemplate a scenario whereby the stress slowly creates, and build-up conditions favoring aggregation and ultimate failure of an organism.

Protein aggregation seems to be governed by the same laws of protein folding and protein-protein transient interaction. At an "extreme" of this continuum we find the polymerization of protein monomers to generate highly structured materials. At the basis of all these phenomena is the possibility to different portions of aminoacid sequence to mutually recognize and interact. The consideration of local periodicity patterns of aminoacid residues along the protein sequence is of paramount importance to judge the aggregation propensity of the relative structure, this was demonstrated at different levels of definition going from the extremely periodic hydrophobicity patterns of polymerizing proteins to the discovery of punctual local peaks of periodicity in globular proteins that have a given probability to give rise to aggregation. More in general the shift between correct folding and multimeric misfolding leading to aggregation can be considered as a basic stochastic process. This stochasticity is probably at the basis of the pathogenesis of all the so called misfolding diseases influencing macroscopic features of these diseases like the slow onset and age dependence. And increasingly there is a growing awareness that given the correct boundary conditions any protein will fold or aggregate. Given the importance placed on oxidative stress for physiological degeneration, it is easy to contemplate a scenario whereby the stress slowly creates, and build-up conditions favoring aggregation and ultimate failure of an organism.

The consideration of local periodicity patterns of amino acid residues along the protein sequence is of paramount importance to judge the aggregation propensity of the relative structure, this was demonstrated at different levels of definition going from the extremely periodic hydrophobicity patterns of polymerizing proteins to the discovery of punctual local peaks of periodicity in globular proteins that have a given probability to give rise to aggregation. In general, the shift between correct folding and multimeric misfolding leading to aggregation can be considered as a basic stochastic process, and of REC). This is perhaps the reason for the TREND based sequence/structure correlations we observe and deserve further investigation to explain a possible "diffusion" process. At the basis of all these phenomena is the possibility to different portions of amino acid sequence to mutually recognize and interact. may be centered around hydrophobicity singularities. This stochasticity may be the basis

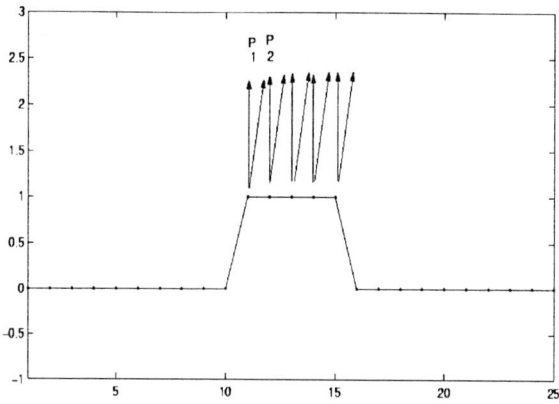

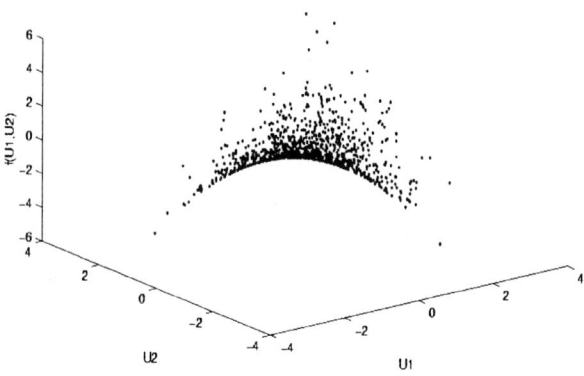

Figure 4.19. A protein hydrophobicity singularity is inherently unstable. Given the particular AA sequence from which it is composed, and the specific boundary conditions (pH, temperature, etc.), it has a given binomial probability associated with it to either 1) fold, or 2) aggregate (top panel). For a given protein then, the singularities form a unique combinatorial stochastic attractor which represents the sum total of its probabilities relative to routes to folding/aggregation (bottom panel).

of the so called misfolding diseases, influencing macroscopic features of these diseases such as slow onset and the age dependence. And increasingly there is a

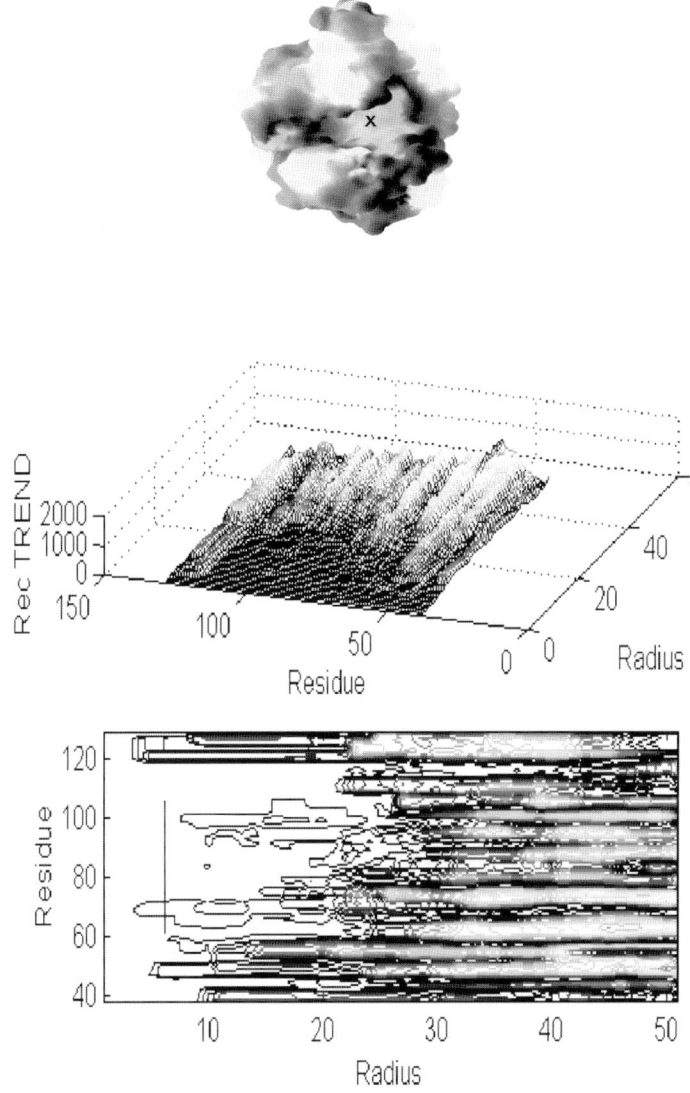

Figure 4.20. Lysozyme with electrostatic coding (top), and "X" indicating significantly negative active cleft. Windowed RQA for lysozyme (bottom). Note the total absence of TREND for residues responsible for cleft (solid bar in contour map) in low radius.

growing awareness that given the correct boundary conditions any protein will fold or aggregate. Given the importance placed on oxidative stress for physiological degeneration, it is easy to contemplate a scenario whereby the stress slowly creates, and builds-up conditions favoring aggregation and ultimate failure of an organism. For example, it has been shown that oxidation can significantly alter protein hydrophobicity. Stated another way, the problem becomes one of combinatorics resolving mathematically in a stochastic attractor based upon the conditional probabilities, and represents the relative routes to folding/aggregation. It is further hypothesized, that since the boundary conditions are dynamic, nonlinearities in the feed back to the conditions can easily develop, and may account for pathology/aging.

In this respect it is possible that electrical dipoles or moments may help explain the findings. This is to recognize that the primary structure of a protein defines the sequence of hydrophilic dipoles, and it is the interaction of the dipoles with the solvent which needs description. This may be the reason for the promising correlations found between sequence-based descriptors in terms of hydrophilicity and three-dimensional structural features (Fig. 4.20). The relative position of these "dipoles" determines protein folding in water. Intrinsic to the concept of dipole is the presence of a direction of some property, the "computational counterpart" of a preferred relative direction between dipoles in RQA terms is the variable TREND that measures the departure from stationarity of the studied series (an approximate, "coarse-grained" derivative of the recurrences).

4.6 Compartment Models

The theory of compartment models represents the basic theoretical framework of biochemical and physiological states in biology and medicine. It is usefully applied to analysis of systems characterized by phenomena of transport and of exchange of biological substances as in metabolic or endocrine systems. A particular metabolic field of application for compartment models is given also by pharmacokinetics and pharmacodynamics. Compartment models are represented by a definite number of compartments (homogeneous pools) being the different compartments characterized by exchange of the considered substance.

Linear compartment models are usually realized symbolically expressing the exchange of matter among the different compartments of the considered physiological system and thus representing the kinetics by a system of linear differential equations that quantitatively describe the exchange of matter and thus the time evolution of the substance considered for each compartment. Mathematical realization of a generic bicompartment model may be represented in the following terms (see Fig. 4.21)

$$\frac{dx_1}{dt} = -k_{11}x_1(t) + k_{12}x_2(t) + f_1(t), \qquad (4.5)$$

$$\frac{dx_2}{dt} = -k_{21}x_1(t) - k_{212}x_2(t) + f_2(t), \qquad (4.6)$$

where

$$k_{11} = k_{01} + k_{21}; \ k_{22} = k_{02} + k_{12}; \qquad (4.7)$$

$$f_1(t) = f_{01}\,\delta(t); \ f_2(t) = f_{02}\,\delta(t); \qquad (4.8)$$

$$x_1(0_+) = f_{01}\,; \ x_2(0_+) = f_{02}. \qquad (4.9)$$

The aim here is to consider a particular case of bicompartment model given by the following set of differential equations:

$$\frac{dx_1}{dt} = k_2 x_2(t), \qquad (4.10)$$

$$\frac{dx_2}{dt} = -k_1 x_1(t), \qquad (4.11)$$

$$x_1(0) = x_{01} \text{ and } x_2(0) = x_{02}. \qquad (4.12)$$

Solving by Laplace transformation one obtains

$$x_1(s) = \frac{x_{01}(s) + k_2 x_{02}}{(s + i\alpha)(s - i\alpha)}, \qquad (4.13)$$

$$x_2(s) = \frac{x_{02}s - k_1 x_{01}}{(s+i\alpha)(s-i\alpha)}, \qquad (4.14)$$

and

$$\alpha = (k_1 k_2)^{1/2}. \qquad (4.15)$$

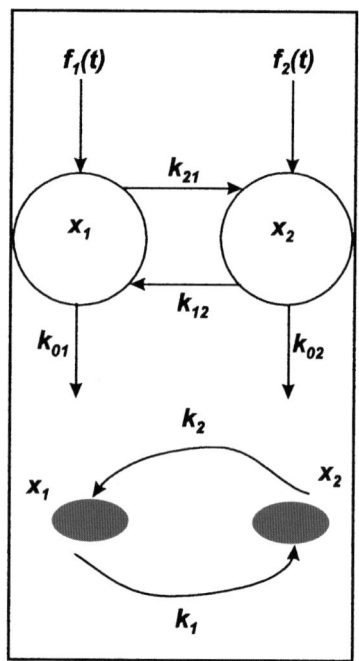

Figure 4.21. Bicompartment model.

Laplace antitransformation gives a solution for the model:

$$x_1(t) = x_{01} \cos \alpha t + \frac{x_{02} k_2}{\alpha} \sin \alpha t, \qquad (4.16)$$

and

$$x_2(t) = x_{02} \cos \alpha t - \frac{x_{01} k_1}{\alpha} \sin \alpha t .\tag{4.17}$$

It is also obtained that

$$x_1^2(t) + x_2^2(t) = x_{01}^2 \left[\cos^2 \alpha t + \frac{k_1^2}{\alpha^2} \sin^2 \alpha t \right] \ldots$$

$$+ x_{02}^2 \left[\cos^2 \alpha t + \frac{k_2^2}{\alpha^2} sen^2 \alpha t \right] + \frac{x_{01} x_{02}}{\alpha} (k_2 - k_1) \sin 2\alpha t .\tag{4.18}$$

For $k_1 \to k_2$, this gives

$$x^2{}_1(t) + x^2{}_2(t) = x_{01}{}^2 + x_{02}{}^2 = R^2 ,\tag{4.19}$$

which are circles in the (x_1, x_2) plane.

The following transformation may be applied now to $x_1(t)$ and $x_2(t)$:

$$x_1(t) \to x_1(t) - A,\tag{4.20}$$

$$x_2(t) \to x_2(t).\tag{4.21}$$

As a consequence of such a transformation

$$\frac{dx_1(t)}{dt} = k_2 x_2(t) ,\tag{4.22}$$

$$\frac{dx_2(t)}{dt} = -k_1 [x_1(t) - A].\tag{4.23}$$

And

$$x_1(s) = \frac{x_{01}s^2 + k_1k_2A + x_{02}k_2s}{s(s+i\alpha)(s-i\alpha)}, \qquad (4.24)$$

$$x_2(s) = \frac{x_{02}s + k_1(A - x_{01})}{(s+i\alpha)(s-i\alpha)}. \qquad (4.25)$$

Laplace antitransformation gives

$$x_1(t) = A - (A - x_{01})\cos\alpha t + \frac{x_{02}k_2}{\alpha}\sin\alpha t, \qquad (4.26)$$

$$x_2(t) = x_{02}\cos\alpha t + \frac{k_1(A - x_{01})}{\alpha}\sin\alpha t. \qquad (4.27)$$

Also

$$x_1^2(t) - 2A\,x_1(t) + x_2^2(t) = (x_{02}\,T - PQ)\sin 2\alpha t + (P^2 + x_{02}^2)\cos^2\alpha t + (Q^2 + T^2)\sin^2\alpha t - A^2. \qquad (4.28)$$

And is reduced to

$$x_1^2(t) - 2A\,x_1(t) + x_2^2(t) = 0. \qquad (4.29)$$

Two significant points may be now outlined. The first point regards the kind of system that may be considered to be connected to the bicompartment model.

It must be emphasized that the equations do not represent a linear compartment model in the traditional sense of this term. The reason is the presence of the term, $k_1\,x_1(t)$ ($k_1 > 0$), that is contained in the system of differential equations It cannot represent an exchange of substance, since it represents instead an acting mechanism of control that the substance x_2 exerts on the substance x_1. In detail, it represents the action of inhibition exerted from x_2 on x_1. Thus, the equations realize a bicompartment model in which the general mechanism of biological control is represented as two different and interacting substances, x_1 and x_2, present in each compartment, x_i (i = 1, 2).

Singularities in Biological Sciences

A significant conclusion is that the equations correspond to bicompartment models, and is one of the most basic mechanism acting in biological matter. $x_1(t)$ and $x_2(t)$ can represent respectively the shift of blood glucose and insulin from the basal level in the condition of fasting for normal subjects. It should be noted that $x_1(t)$ and $x_2(t)$, represent the behaviors of two coupled oscillators

If the special case bicompartment model equations are combined, it is easy to show that

$$\frac{d^2 x_1}{dt^2} + \alpha^2 x_1 = 0, \qquad (4.30)$$

and

$$\frac{d^2 x_2}{dt^2} + \alpha^2 x_2 = 0, \qquad (4.31)$$

and, combining with the transformed equations

$$\frac{d^2 x_1}{dt^2} + \alpha^2 x_1 = \alpha^2 A, \qquad (4.32)$$

and

$$\frac{d^2 x_2}{dt^2} + \alpha^2 x_2 = 0. \qquad (4.33)$$

Careful inspection of these equations reveal that they are of the basic form as those previously suggested by Dixon (see previous chapter). They represent a deterministic system given by two coupled oscillators with every point $[x_1(t), x_2(t)]$ belonging to a unique solution described by a circle with radius $R = (x_{01}^2 + x_{02}^2)^{1/2}$.

It can be shown that

$$A = \frac{x_1^2(t) + x_2^2(t)}{2x_1(t)}, \qquad (4.34)$$

with a family of transformed circles all sharing a common tangent point at the origin. The significant point is that such a common point is intersected in a finite time, and, in addition, it does not represent a fixed point. There is a singularity at the origin, a violation of Lipschitz conditions. Here neither past nor future time evolution of the system are uniquely determined. Determinism fails for such a system.

In conclusion, the dynamical system represented by these equations is a nondeterministic system made by two coupled oscillators. While the physical state of such a system is far from the (0,0) point in phase space, external noise will not affect it, provided that the average amplitude of the fluctuations is small compared to A for this trajectory. Instead, as the trajectory approaches the origin, noise will play an essential role. Solutions for different A converge together and all intersecting in (0,0). Noise will cause the trajectory to jump between solutions of widely different A in a random way, resulting in a piecewise deterministic dynamics and, a nondeterministic chaotic behavior.

Comparison of the models with real data demonstrate the plausibility of the alternative view. The results are given in Figs. 4.22 and 4.23. It is seen that, despite the roughness of the employed bicompartment model which exhibits remarkable approximations of glucose and insulin utilization , an evident and satisfactory resemblance is obtained between predicted and experimental data. Using the bicompartment models, a demonstration of a simple case of compartment theory to suggest that nondeterministic chaotic behavior may have an important role regarding fundamental mechanisms of biological matter has been presented. In order to reach this objective it must be emphasized that the kind of biological mechanism does not represent linear compartment models in the traditional sense of this term. There are some specific reasons in this regard. Compartment models are usually expressed by differential equations representing the exchange and the balance of the mass of a given substance among the different compartments. Instead, the model presents the case of two coupled oscillators.

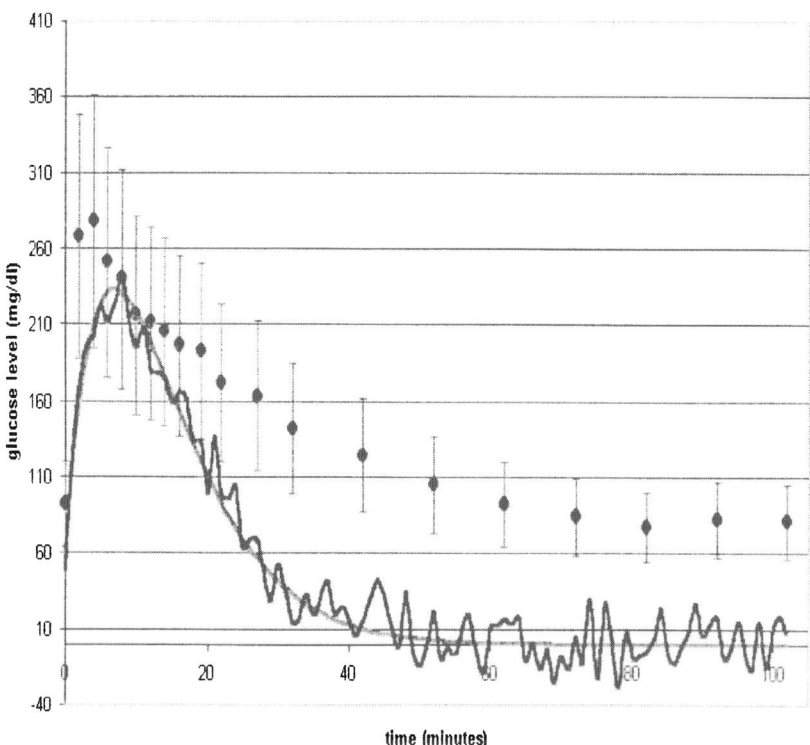

Figure 4.22. Fitted experimental data compared to bicompartmental model. Circles = experimental data; solid line = fitted model; jagged line = noise perturbed model.

A traditional approach to physiological regulation has prevailed based on the general recognition of the central role of negative feedback loops in system dynamics. Feedback interpretations represents the reaffirmation of the validity of basic principles of science as determinism and cause-effect dynamics. However, experimental evidence suggests that in biochemical and physiological observables, the required adjustments may be realized also before the considered controller has had the possibility of experiencing feedback. In addition, no rule

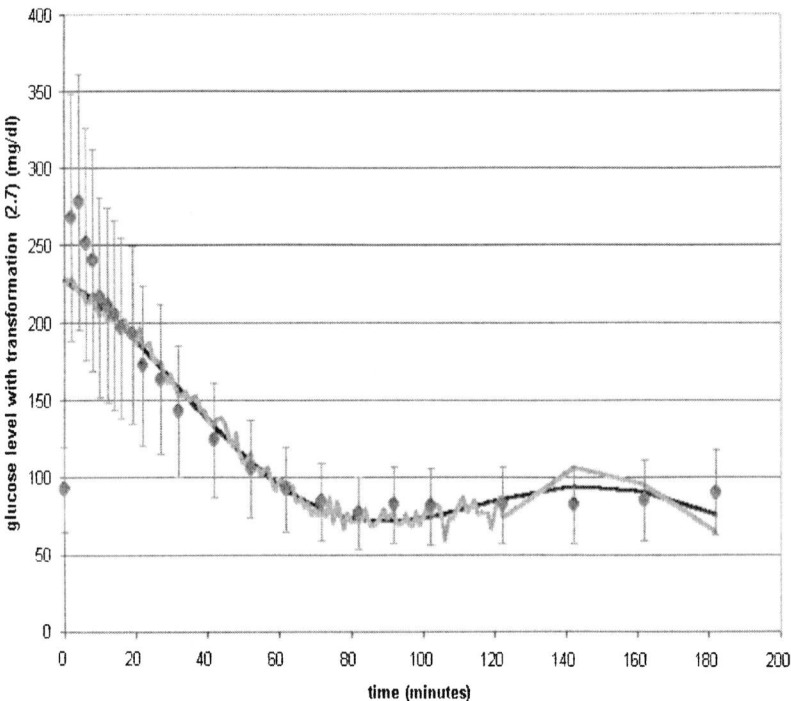

Figure 4.23. Fitted experimental data compared to transformed model bicompartmental model. Circles = experimental data; solid line = fitted model; jagged line = noise perturbed model.

seems definitively written that nature must preserve always a general and absolute Lipschitzian behavior. The standard approach of classical dynamics assumes an absolute Laplacian behavior for natural dynamics of systems. In this vision of physical and biological reality, time evolution of a system is retained to be uniquely determined by its initial conditions. All the deterministic visions of nature and, in particular, of biological dynamics holds to the uncriticized assumption that physical and biological systems must be Lipschitz without exceptions. Instead, as has been shown, the basic theoretical framework of biology, as it is represented from compartment theory, exhibits cases of compartment models with non-Lipschitz equations of motion. The consequence of such a violation is the possibility of non-unique solutions. If a dynamical system is non-Lipschitz at some singular point, it is possible that several solutions will intersect at this point. Since a singularity constitutes a common point among many trajectories, the dynamics of the system, after that the singular point has been intersected, is no longer determined from previous system dynamics. Thus a profound and radical change is evidenced in the dynamics of such a system: it no longer exhibits a deterministic but a nondeterministic dynamics. One possible consequence is nondeterministic chaos. In fact, in the considered nondeterministic system the various solutions move away from the singularity and they tend to diverge. Several solutions coincide at the non-Lipschitz singularity and whenever a phase space trajectory comes near this point, any arbitrary small perturbation may drive the trajectory to a completely different solution. Since noise and/or deterministic chaotic disturbances may pertain inevitably to any physical or biological system, the time evolution of a nondeterministic dynamical system will consist in a series of transient trajectories with a new one selected randomly whenever the solution approaches the non-Lipschitz point in the presence of such noise or deterministic chaotic perturbations. As a result, alternative to the consolidated principles of determinism and of consequent deterministic chaos, such systems could exhibit also nondeterministic chaos. This is precisely what has been shown to be possible for the compartment models discussed.

4.7 Biological Complexity

In sum, it is proposed that there are at least two types of chaos operative within biological systems: deterministic and non-Lipschitz. Recognition of this allows for a resolution to the quandary regarding observed physiological processes which require adaptability as well as many degrees of freedom for their adequate description.

The coexistence of these two forms of dynamics helps also to elucidate the debate between "order and complexity": it has been pointed out that statements regarding biological existence as being "highly ordered and complex" are non sequiturs. Complex systems are described by high entropic values; whereas ordered systems have low entropy. And yet intuitively, scientists have noted that this does not adequately describe the situation in biological sciences: there are obvious highly ordered structures—yet, their function or activity is often impossible to predict precisely because of their complexity (high entropic value). Non-Lipschitz dynamics can accommodate both: the order resides in the piecewise determinism of trajectories, but the complexity resides at the singular points. Here the probabilistic nature of the trajectories preclude deterministic prediction in contradistinction to deterministic chaos which in theory can be predicted given infinite precision. Again, material randomness presents itself as an important entity in the generation of life, and puts into question statements regarding functional order out of chaos. In some sense, it may be said that things evolve and adapt to the environment by injections of bits of random noise, otherwise they get stuck into the dullness of mathematics. Is this a limit of our mathematics (after all differential equations are not the absolute perfection) or a necessity of life?

This view may be bolstered by studies where we compared the "randomness" of mathematics (random number generators, the transcendental number "pi," and radioactive decay): the radioactive "randomness" was determined to be more "random." This was done by evaluating the dimension at which any signs of determinism for a given kind of randomness failed to appear. The rationale is that truly "random" series fail to exhibit values of the RQA variable, DET no matter how high it is embedded (Fig. 4.24). On the contrary, non-random series produce values for DET in higher embeddings. This is an important consideration when dealing with computer-simulated systems. So-called randomness developing from such models, or the random number generators used are not truly comparable to randomness from physical systems. Although it is frequently suggested that although this may be so, from a phenomenologically qualitative perspective the difference does not exist. It is not clear how justification for such an observation can be made: if there is a true difference between physical randomness and simulated randomness, the sequelae from the model may be totally different, and predictions may indeed be suspect. A common example is that of weather forecasts. Some of the world's most powerful computers are used to make climatological predictions, yet they can do so only for a few days into the future with progressively increasing uncertainty. The explanation for this is that models may be not sufficiently refined, or there is insufficient hard data input. Yet there is no doubt that these models are reasonably successful given the enormity of the problem.

The biggest challenge will remain to "model" the stochastic components involved in prediction. A common practice is to add stochastic terms from various noise "distributions." Yet, if the dynamics is constantly changing, one must consider the fact that the noise distribution may be changing. Indeed, one must consider, given the possibility that there really is a difference between real and simulated noise, that physical noise has no permanent distribution. It is interesting in itself, that often, depending upon the phenomenon being observed, different noise distributions may be appropriate (Fig. 4.25). The importance of this topic can be appreciated from the perspective of "rare events." Significant climatological catastrophes such as flooding can impact populations to an enormous degree—yet the probabilities involving these events are poorly understood.

On the biological level, a similar discourse can be obtained. Investigators are beginning to appreciate the influence stochastics has on basic physiological processes. The initial euphoria regarding the attainment of genomic maps of DNA has gradually given way to the realization that the dynamics involved present significant possibilities for the influence of random events, as the noted DNA entrepreneur J Craig Venter has asserted. Thus the traditional view of a "homeostasis" becomes a contradiction, and the arguments regarding the importance of "nature vs. nurture" appear to be poorly posed. The need to frame the discourse in terms of dichotomies is most likely ill-posed, since there is a connotation that there is a "real" reality constantly being altered by some unwanted factors. As I have argued, instead, the discourse should be approached from the view that these other factors are part of the dynamics themselves: they are part of the evolutionary process on many length scales. As such, as been indicated, the biochemistry at the heart of biology is a discrete event process which does not necessarily tie itself to the past. This is the process of natural creativity which some call evolution.

If biology involves singular dynamics, the entire approach to what constitutes what is normal and abnormal needs to be reevaluated. If the "phenotype" is part of a grand process depending upon layers of probabilities, are "abnormal" phenotypes simply defined by the rarer possibilities? Can "engineering" approaches to management of these possibilities be successful?

It would seem that these questions cannot be easily addressed until the overall paradigm to such problems is altered to some degree. At its most fundamental level, the mathematics of biological scientific inquiry must admit more legitimacy for statistics and probability. Traditionally, statistics have been relegated to epidemiological phenomena. It would seem, however, that the vagaries of singular processes should allow for a legitimate role for statistical approaches. As has been outlined here, these statistics need not be in the traditional sense, and are integral to the dynamics themselves, and can serve as a foundation for an approach which avoids the Procrustean bed of artificial "logical" assumptions.

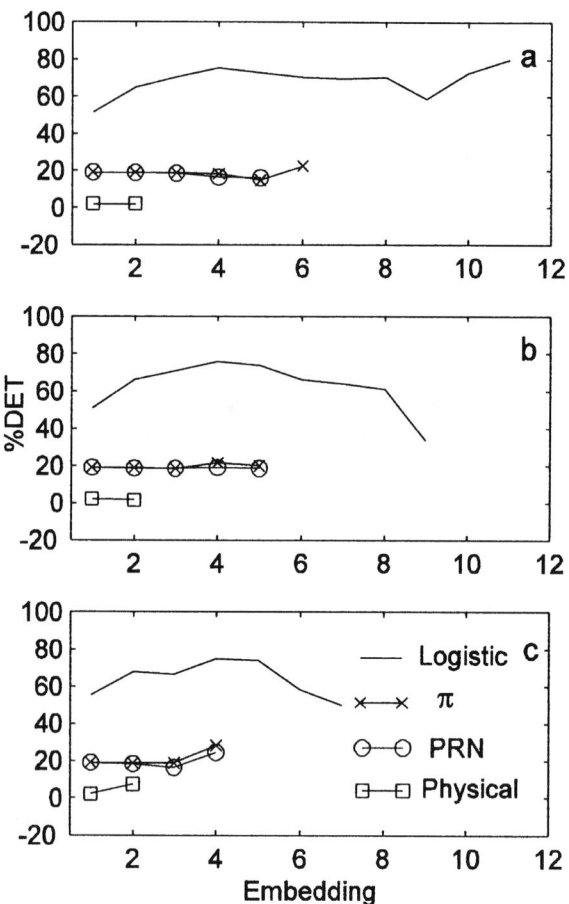

Figure 4.24. The effect of progressive embedding upon the DET variable for progressively longer (a = 512; b = 1024; c = 2048) "random-like" number series. Note that lack of DET begins at an embedding of 2 for a physical (radioactivity) series. (Logistic = series from the logistic equation; π = calculation for progression decimal number places; PRN = pseudo-random computer generated series.

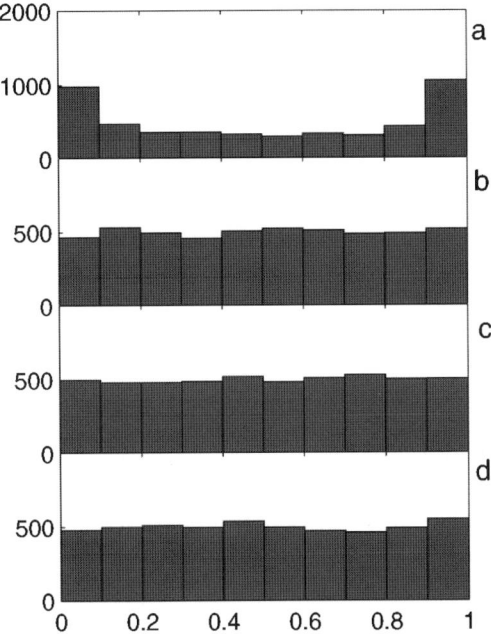

Figure 4.25. Histograms from the random number study of the previous figure (a = logistic, b = pi, c = pseudo-random, d = physical). Note the similarity of the distribution (uniform ?) except for the U-shaped of the logistic.

Borrowing from David Mumford, it should be pointed out that the mathematics of formal logic upon which traditional mathematical models are designed is only one form of abstraction. Psychology, on the other hand points out that we make decisions based upon a weighing the likelihood of possible events through the probability of posterior events. A formal logic does not exist in practice; rather a quasi "Bayesian" analysis. In effect, as I have argued, this involves a stochastic attractor of brain events and memories which is totally dependent upon the combinatorics of the "net." If science is part of this experience, explicit recognition of this process allows for a better understanding of the goals of science. To some degree, implicit in this is the recognition of Karl Popper's protestations regarding "falsification": however many confirming instances there are for a theory, it only takes one counter observation to falsify it—only one black swan is needed to controvert the theory that all swans are white. This is to say that observations are essentially probabilistic—contingent upon the vagaries of time.

These statements recall some of the ideas presented in the Introduction; namely, that the perception of scientific principles is not what scientists actually do. The public persona of science is a time honored canon of ideas which embody the logic of testing hypotheses based on objective observation. There is the idea that the scientific endeavor is a cut and dried, mechanistic process designed to obviate personal bias. To a degree it does—but as Mumford and Popper have suggested, this may not be the whole truth. At the same time anecdotes and reports suggest that many scientists do not develop strict hypotheses, but instead often depend upon "hunches" or "intuition," which are nevertheless subjected to experimental verification.

The idea that science is "logical" imposes an Aristotelian/Platonic preconceived world view on observations which is biased in itself. To say that the world is "logical" is to some degree incorrect. Indeed, if it were logical, there would be no need for science in the first place: this was the idea behind many medieval philosophers' discourses suggesting that the world's functioning could be discerned from logic alone. Lovejoy's influential tome on the "great chain of being" clearly presents a world view which saw everything in reality as part of a grand scheme that places everything into a proper place according to a categorical order. And the idea brought a certain comfort to the medieval mind to know that although surroundings might appear chaotic, there was indeed a certain proper order governing things.

Although Hume and Bacon (Francis and even Roger) began to break down this assumption, the "logic" aspect persists. Interestingly enough, it has been the experimental psychologists who have provided the evidence that logic is a construct, as much as any other construct. Cognitive scientists have demonstrated that thinking and the acquisition of knowledge is far from logical, and is perhaps best described as a progressive process involving acquisitions (perceptions) which are then tested via "prior probabilities" to affirm or deny some new perception—much of which develops on an unconscious level. In some ways this is a kind of scientific process all of its own, requiring the continual testing of new observations.

What is important to note about these developments is the fact that certain sectors of society have taken these findings to suggest that reality as such does not exist, and instead is a creation of the mind. As a result, they view the scientific proposition as being part of a scheme which allows them to challenge science on its own noetic grounds. This attitude has been especially noticeable among some circles of existentialists and post-modernists.

To some degree their view is correct: the mind does develop certain perceptions and "histories." Yet it is science's strength that it tests these perceptions. The history of science demonstrates this constant struggle to verify its ideas. And insofar as scientists are humans, it really should not be a surprise that human motivation sometimes intrudes on objectivity.

By suggesting the possibility of nondeterministic dynamics, I submit that such a paradigm allows us to more easily understand the objects of scientific endeavor as well as our role within it. The allowance of probability a greater role in the paradigm of dynamics, does not dismiss totally the idea of determinism, but rather clarifies it as a "piecewise" determinism. And reality becomes a part of a Bayesian process to clarify objective observation.

5 *Singularities in Social Science/Arts*

I have been often asked to comment on the applicability of our theory of noise and singularities to the arts and social sciences. I have done so with some concern and caution, insofar as these areas tend to be more difficult to evaluate since when compared to the physical and biological sciences, objective data are difficult to come by. Nonetheless, I have tentatively ventured in these areas so as to provide cautions regarding unrestrained metaphorical and qualitative appropriations of these ideas as has been seen in the broad area of "chaos theory." (I may also add to this, the area of "catastrophe theory," which has tended to remain a qualitative field, from which I distinguish our ideas in that I emphasize objective measures.) As I have stressed, care must be had to find objective data to support the idea of singularities and instability. This is obviously difficult when dealing with many degrees of freedom systems such as are those in economics, art (from the viewpoint of perception and cognition) psychology and sociology. The arts especially are difficult to evaluate given the immense importance of the subjective experience ("de gustibus non est disputandum"). I am heartened, nevertheless, by the fact that this subject has been broached by the noted Roger Penrose.

A cursory view of these disciplines, however, reveal some possibilities which specialists may entertain. Specifically, in economics, the ability to obtain data has been greatly amplified with the advent of more powerful, easily obtained computers. In psychology, cognitive processes are being accessed with the help of PET scanners, and functional MRI. Insofar as visual arts are also to some degree cognitive processes, they may similarly be evaluated (although, again, because of their subjective, and often cultural nature, conclusions must be carefully limited). And the social sciences have greatly refined many of their tools again through better data collection as well as tools to help unravel multiple degrees of freedom. In sum, there appear to be some possibilities, but cautious, objective analysis must be performed first. In what follows are some possible directions.

5.1 Economic Time Series

Considerable efforts have been expended to determine if economic time series can be modeled as nonlinear chaotic dynamics. A motivation has been the recognition that often, economic observables appear chaotic, although they are clearly generated by some form of determinism. Unfortunately, these attempts have not been distinguished by unequivocal results. This follows the results of other sciences such as biology. Whereas the physical sciences can generate reasonably long stationary data, economic sciences perforce depend upon imperfectly collected sources. Additionally, greater amounts of noise often attend such data. Some of the difficulties have been attributed to the algorithms used to quantify chaotic invariants, such as dimensions, entropies, and Liapunov exponents. Other, more traditional methods, such as FFTs have also exhibited well-known difficulties. As a result, much energy has been devoted to analyzing various methods of calculation and conditioning to real life data.

In the case of economic data, a fundamental problem has not been addressed; namely, the correspondence of economic data to the fundamental assumptions of chaos and other fully determined models. Specifically, it is noted that chaotic dynamics are continuous, deterministic systems. Although the time series of such systems are random looking, processes which generate them are not: their behavior is, by definition, rigid. Economic systems, on the other hand are subject to the vagaries of human agency. Although these facts are obvious, the consequences are seldom appreciated. To suggest that economic data is chaotic, would assume that such systems are ultimately very unstable. This comes as a result from the recognition that the way to change a chaotic system is to change its control variable or by creating specific perturbations, but, as is known, very small changes can produce significantly different behavior. Economics is a system or an historical trajectory thru different systems. Probably economics has its "laws" but they hold only for relatively short periods between a "transition point" and the subsequent (drift) in which economy moves among different versions of the same "laws."

5.1.1 Stock Market Indexes

As has been done previously, RQA was used to examine some economic data. RQA of a data series can begin in a number of ways, but one useful approach is to view the global recurrence plot. Frequently, this can identify some basic dynamics relative to multiplicative (intercorrelated vs. independent) processes. I have taken the stock index of the New York S&P 500 from January 6, 1986 until December 31, 1997, and performed such an analysis.

- If one looks at the recurrence plot, two things are apparent (Fig. 5.1):

- the data are not stationary, given the paling of the recurrences (i.e., they become less frequent); and

- within the nonstationary processes, there are several embedded stationary processes characterized by square structures, while the over-all funnel shape is typical of a transient.

What is more interesting is the numerous box-like structures typical of local autocorrelated data. Furthermore many of these structures have additional "boxes" within them. These typically terminate abruptly, only to have another begin (Fig. 5.1, arrow). Thus one can conclude that multiplicative (intercorrelated) processes are present. If we take a histogram of recurrences (in lagged order), this impression is correct (Fig. 5.2). A scaling region can easily be identified. Scaling suggests a multiplicative process, while the horizontal regions suggest independent "noise-like" processes. The scaling region, however, is bounded by two non-scaling regions: the first is relatively horizontal, and the second has several peaks. Being quite fast, and of relatively equal probability, this first non-scaling region suggests noise of some type. The second region has several peaks. The end of the scaling region ends at 95 days and may suggest some quarterly process since the peak immediately to the right is about 125 days. Similarly, the next cluster is centered around 350 days and may be part of yearly cycles. However an area immediately adjacent near 500 is not clear as to its source.

Figure 5.1. Recurrence plot of stock market index. Note the several box-like structures (arrow). Within these are even more such structures suggesting a transient multiplicative process.

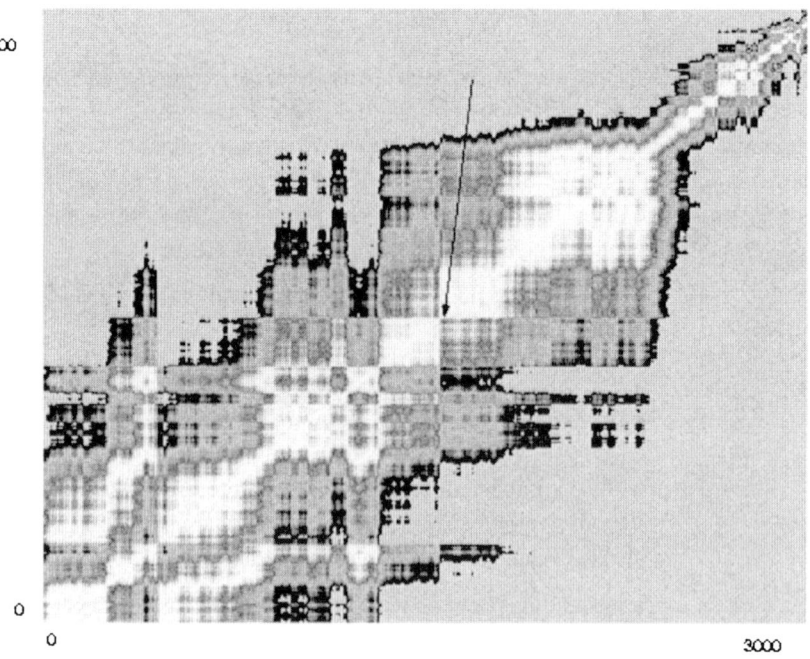

To obtain more information, a windowed RQA was done (Fig. 5.3). Using a 90 day window, it becomes clear that there is considerable variation in the series, but the RQA variables provide some hint as to their significance. Although there is considerable change in relative magnitude of the index, the DET and the Liapunov approximation provide the degree of importance (see Fig. 5.3).

Several regions are numbered in the time series of the Index. A noticeable precipitous fall (1) is presaged by a minor decrease in DET (2a) and an increase in the Liapunov exponent. What is more noticeable is the area (2). Although the time series appears rather unremarkable except for its gradual trend upward, the DET (2a) and Liapunov (2b) are considerably unstable, which is immediately followed by a series increasing in its slope upward (3). (Instability was determined by calculating a 30-day sliding window of 95% confidence limits. Variables falling outside the limits were considered unstable.) Note, however, that the DET (3a) and Liapunov (3b) are relatively stable, while the TREND clearly tracks variability. Thus, although this region is changing—it is a "stable" change, given that the DET and Liapunov remain relatively constant, suggesting an underlying stationary process.

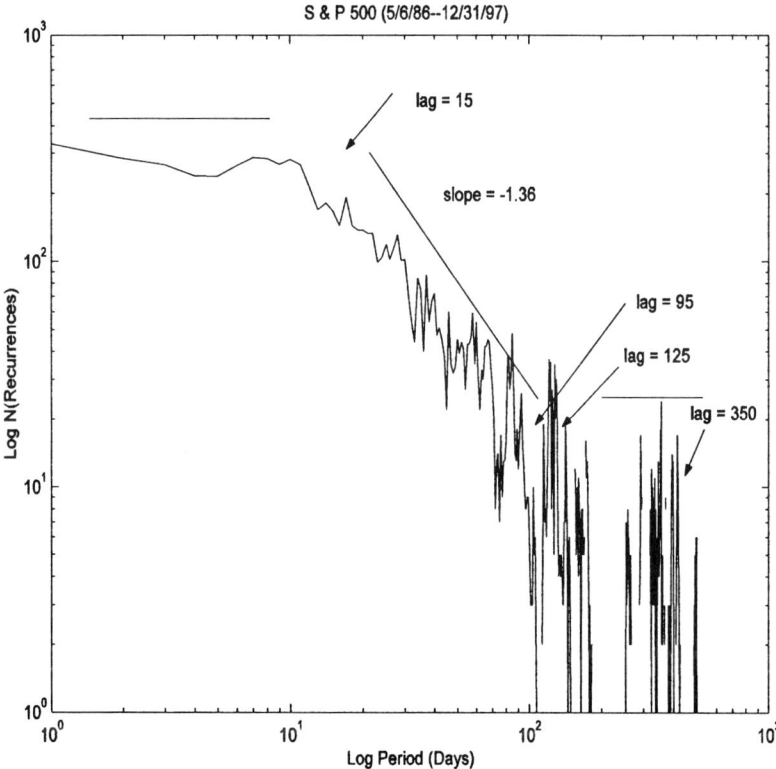

Figure 5.2. Histogram of recurrences. Individual peaks allows for comparison with the actual data time series.

It appears obvious that a typical stock exchange index is quite complex, and although it may have features typical of stochastic processes, it also has areas of relatively significant determinism. To try and make sense of the process, I suggest that financial markets are characterized by singularities of the discrete data. At the singularities, the dynamics are very sensitive to "outside" data, which may force the series to different regimes, not necessarily related to the previous history.

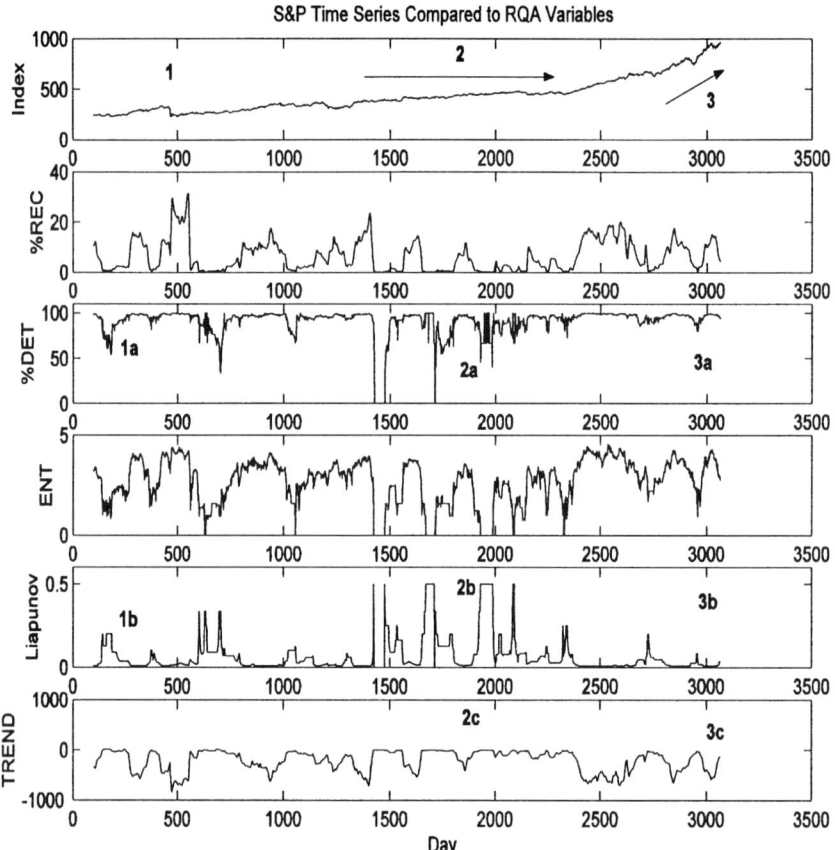

Figure 5.3. Sliding window RQA analysis.

One additional note: although there are many models which add stochastic terms, these terms are added to the entire dynamic. The model here proposed would suggest that the stochastic aspects appear only at the singular points, and thus do not distort the general dynamic of the trajectory. The difficulty, of course, is to determine the proper form of the deterministic "trajectories" and their related probability distribution functions. At each singular point, a variety of connecting possibilities are admitted—each with their own likelihood of being realized. Furthermore, the strength and type of excitation conditions these possibilities. Thus, the final problem is cast in the form of a combinatorial problem, resulting in a "stochastic attractor."

5.1.2 *Exchange Rates*

Exchange rates for the European Union compared to different currencies were examined. Examination of recurrence plots of the series demonstrated total lack of stationarity. Further evaluation by the use of surrogacy techniques for hypothesis testing as a result was not possible.

Recurrence plots were generated for several different currencies This resulted in a topologically qualitative classification of patterns: constant recurrence plot over time, high recurrence during the first 6 months examined or in the last 6 months examined, and an "airplane-like" structure. It was clear that some time series were on the borderline of these classification schema and to classify them in one group or in another was rather subjective. Instead, individual analysis of high frequency currency exchange rates from 1996 using RQA allowed us to establish a correlation between changes in the system and external factors. The fact that there was considerable variation in the measure of DET during the year may be indicative of changes in forecasting potential. Specifically, the idea was entertained that such a qualitative dynamical change might be related to a unique event. That is an unstable singularity may have dislocated the processes normally operative This possibility was suggested by a plot of REC points: this resulted in the demonstrations of two regions clearly correlated separated by an amorphous region probably related to the unstablizing event. may be related to singular events, which effect an important change in the dynamics (Fig. 5.4).

5.2 *Art (The Science of Art?)*

It has been suggested that what separates a true "artist" from a craftsman is the artist's ability to create the element of surprise in the observer. Certainly poetic and philosophically interesting, this statement is difficult to objectively evaluate (What is the definition of "surprise," "artist"?). However, there may be a way to suggest a possibility in terms of visual and cognitive function in conjunction with the notion of singularities. (Clearly my experience is not in this

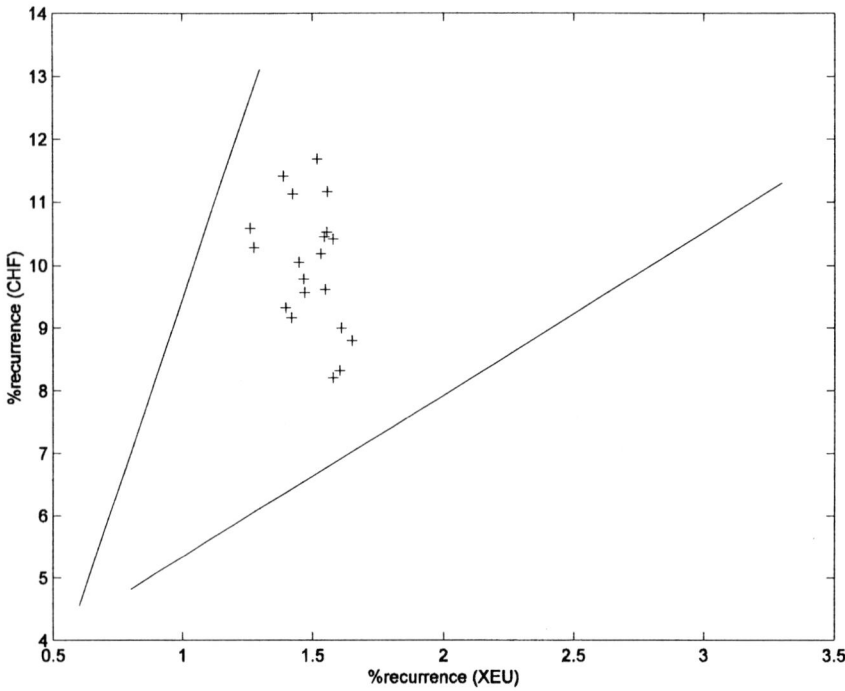

Figure 5.4. Plot of the REC for the ECU and the Swiss Franc. Left line for the period between January and March; a transition period in between the two lines indicated by "+" points; and right line for the last part of the year.

realm, however, I suspect that experimental psychologists and neuroscientists— especially with newer tools such as PET scanners, etc.—may be able to devise experiments to pursue this view.)

Certainly what constitutes "art" is a very tendentious topic. It is confounded not only by its definition, but also by various philosophically and politically motivated disputations: is art for art's sake, or does it need to serve some specific purpose? Nineteenth century continental salons disputed such questions with near fanatical fervor. No doubt both views can be strongly supported, but one thing seems to be clear, as Mihaly Csikszentmihalyi has pointed out, genius is not genius unless it is publicly recognized. Stated another way, the social context of artistic endeavor is a given. Art is a form of expression and communication, and if there is no recognition (communication), no matter how remarkable it is, it has not served its purpose. In some sense this view goes contrary to the assertion that art is a form of self-expression. This may be true, but I suspect this is more a function of the popular tendency to label all forms of expression as art. What I have in mind is the public recognition of the "surprise" I mentioned previously. (Again, as a working definition, if it is not communicated, it remains self-expression.)

5.2.1 Examples

I present two well-known public pieces of art and suggest that they are remarkable because their perception is based upon singularities of visual perception, which are perturbed by "noise" in the form of asymmetries. (Many other examples could be used, however, copyright laws limit their adequate discussion.)

The first are the well-known Spanish Steps in Rome (Fig. 5.5). They are well-appreciated for their Baroque execution, but what is also remarked is the sense of "motion" they convey. In looking at the original plan, it is clear that the steps are conceived around a slight asymmetry (Fig. 5.6). Thus the observer seems to be forced to compare sides—a kind of noise, and the comparison does not have a "unique solution" (otherwise it should be simply another form of symmetry with a different shape with respect to central symmetry). Using physical terms this corresponds to saying a system is "frustrated," i.e., has a multiplicity of "quasi-ground states" that are the essence of surprise, because no solution is the definitive one. The viewer cannot "settle" on a perspective. If this asymmetry were greater it might have engendered an unpleasant uneasiness. Thus the effect is based on a minimal change.

A more contemporary example is that of the Chicago Picasso statue which was presented as a gift to the City of Chicago, and resides in the Daley Plaza. A view from the front demonstrates its asymmetrical nature (Fig. 5.7), which again forces the observer to move to view the statue from different perspectives. (Many have interpreted the statue as resembling an Afghan hound.) Indeed, a view from the side, however, suggests the profile of a woman or man (Fig. 5.8). Clearly again, a slight asymmetry at a singular point forces changes in perception. This may be from "cognitive dissonance;" nevertheless it should be noted that the asymmetry is subtle. Because the asymmetry is subtle, the resultant "surprise" is all the more satisfying. A more heavy-handed approach might defeat the effect.

Figure 5.5. Spanish Steps in Rome. It has often been commented that they give the sense of motion. Note that the figure is purposefully outlined in black to emphasize the asymmetry.

Both of these examples recall the theory of protein folding along a possibly frustrated "energy landscape": in this view, the protein seeks a certain ground state of its energy. However, the landscape is not smooth, but rough, which forces the protein to get "stuck" here and again in little valleys, only to get out, and be forced in another "depression."

What has not been addressed, though, is the cognitive aspect. Does the "singularity" of perception induce a "singularity" of cognition? This, of course makes the topic more complicated, but I would suggest that such an increase in the "complexity" of the issue models the significance of the "art," and allows for the inclusion of other forms of art such as writing, poetry, and music.

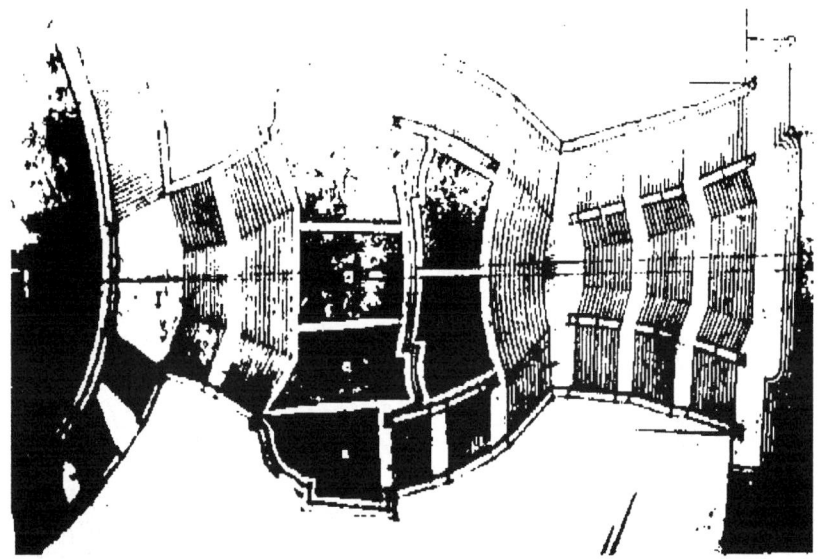

Figure 5.6. Floor plan of Spanish Steps (The Piazza is located at the right). Note the slight asymmetry. (Figure is purposefully outlined in black and white.)

As an exemplar, consider a novel. There are many such works created every year, and many find themselves on best seller lists. Yet most informed readers can easily tell the difference between well-crafted, interesting works, and those which are perhaps more significant. Again craftsmanship does not always equate with "art." At the time that "The Man from La Mancha" was conceived, many popular novels dealt with chivalric themes, but it is only Cervantes' work which appears to be remembered and continues to be an inspiration. Similarly, many detective novels were being produced in 19th century Russia, but only Doestoevsky's "The Idiot," or "The Brothers Karamazov" persist. Similar arguments can be made regarding cinema: how does a work of art force the knowledge of a wide range of thought without becoming so uncontrolled as to be meaningless?

154 Unstable Singularities

Figure 5.7. Picasso sculpture in Chicago. Again a slight asymmetry is noted. The figure is outlined with de-emphasis of background elements.

Music is a much more difficult topic to address. Aural perceptions are transient, but they result in similar brain processes qua perception. They can induce thoughts, emotions and memories. And again, it is suggested that the surprise element of "singularities" define the great from the mediocre.

Figure 5.8. Picasso statue viewed from behind at an approximate 45 degree angle. Presented in black and white outline.

An additional element is the well documented fact that "artists" typically use portions of the brain not commonly used by the vast majority of people. Does this make them an elite? Not necessarily, for invoking Csikszentmihalyi again, it is their "responsibility" to make their craft accessible—obscurantism does not communicate.

Finally, it has been observed that in various forms of perception/cognition EEGs have been described as being "coherent" in specific regions of the brain. The term "coherent" must be taken with some informed consideration: to speak of a dynamic which is constantly changing and dependent upon the natural noise of synaptic connection may give a false impression. As I have argued previously, the brain works to form its "concepts" through stochastic attractors; i.e., approximate attractors which can create a rich variety of similar connections. Furthermore, when considering the compartmentalization of the brain, including right and left hemispheres, the notion of singularities can be traced on multiple levels of organization. Thus the important consideration is not the coherence itself, but the way it is formed: are the synapses rich, or are they narrowed? (Does a good "technical" work fail to be "art" because it is stereotypical—a failure of surprise?) A narrowed probability of choices approaching a determinism suggest a pathology; i.e., an inability to adapt. Could this, through the agency of deranged neurotransmitters be the basis of psychological pathologies which "drive" people to certain stereotypical activity? Are there hierarchies of motivation perhaps intertwined with and condition by, e.g., hormones, peptides?

In this larger context, "coherent" patterns in stochastic attractors, as I have argued, form the basis for the sense of self via one's history. (A person who has lost all memory typically does not have a sense of self.) Depending upon the kind of memory present, a person can suffer from what may be considered the "curse of historicism"—the need to constantly refer to ones past so as to be unable to deal with the present—a theme frequented by existentialists in art and science. (I use the term "historicism" to refer to the unique personal history, which may not always conform to objective facts, and represents a compendium of the perceptions of history.)

In a general sense, it would seem that "art" (be it visual, music or written) is able to induce stochastic classes which are relatively diverse (yet connected), but not necessarily being meaningless. (Are dreams random connections?). In other words, there would seem to be a very distinct boundary between complete "randomness" and "focus" relative to the "controlling" function of the artist.

Finally, a brief comment regarding art "education" should be made: often it is suggested that popular appreciation of the arts is dependent upon a proper education with respect to the art in question. This should be understood in terms of the world view of the perceiver. As has been suggested, we are to an extent our history, and if that history does not include an understanding of perhaps very different social contexts of older works of art, the desired "surprise" may not obtain. Common, contemporaneous understandings are not there, and "communication" is not present. This becomes a delicate question, since it can affect research choices in this area. Are some works of art really so (in the terms I have been presenting), or are they really part of a received view which says more about privileged historical knowledge? Again, referring to Csikszentmihalyi's idea that creativity must be communicated to be considered remarkable, judgment by a privileged elite does not necessarily constitute "art." If anything, privileged knowledge may only confer the idea that some "art" is really remarkable technology (which is not to denigrate the work). Remarkable technology can be justifiably considered in high regard, but should not be confused with the surprise of art. (Consider "Mozart and Salieri", as presented by the great Russian poet, Pushkin. The drama centers around the problems of reconciling "divine" inspiration with the "mundane" existence of everyday life.)

This difficulty assumes a slightly different flavor when dealing with contemporary or "modern" art. It is frequently heard from popular circles that they do not understand such art. Professional critics find themselves duty-bound to "explain" the art. (One recalls the story of an academic specializing in the poetry of a famous South American. He devoted his entire professional life explaining the poet's peculiar symbolism. Upon meeting the poet in person, he asked him about a peculiar phrase in one of his poems relative to its meaning. The poet responded with some surprise and remarked that apparently the phrase was part of a typographical error.) The problem would seem to be dependent upon the view that artists view things in different ways which may be disconcerting to contemporaries. In essence, the artist must in some way divorce himself from historic, traditional views, while not at the same time totally abandoning them—again the problem of maintaining communication. This may require that some time must pass before contemporaries can "integrate" the view with their own "histories." In this sense, it is remarked that the Picasso statue was at first poorly received by Chicago's citizens, but has subsequently been embraced by them. If a work requires so much elaboration, or never really becomes appreciated, one questions if the work is really art—it fails to communicate. Canonization by a select few may not meet the test of communication. Thus it is the "curse of historicism"—between people and/or generations that is really the challenge of the artist. The artist must overcome the "inertia of the past" while at the same time looking to new formulations which do not totally divorce the observer from the observer's own history. No doubt this is one of the reasons that artists are frequently judged as being unique or peculiar. Uniqueness, of course, does not mean that something is remarkable art.

Certainly this recalls the old philosophical distinction between *form* and *matter*, or style and substance. Philosophers have seemed to consistently make a distinction between the way things appear vs. what the things really are. Often, it is remarked that substance is the important thing, and that style by itself is specious. Perhaps this dichotomy is overdone. No doubt substance is important, however, if the substance is not presented in a manner that engenders communication, it would seem to be ultimately fatuous. This recalls the old Biblical injunction about "hiding a lamp under a bushel basket." The communicative aspect of art (as well as perhaps interpersonal communication, of which art is a part) requires and attention to all aspects—perspective as well as cognitive.

5.3 *Psychology*

I next turn to matters of cognition associated with perception: just as a singular point in art perception can force changes in the way things are viewed, so too can ideas. And indeed, what has been said regarding the perception of art can be applied here as well. In a more general sense, an interesting example can be seen in the history of science. Certainly it also applies to perceptions of not only history in general, but also to personal history. Again, I refer to this peculiarity of historical perception as the "curse of historicism." Before Copernicus, for some 500 years, the Ptolemaic view of a heliocentric universe was universally accepted. One obvious reason for this is that, in fact, it fairly accurately predicted the motions of heavenly bodies. It was only, however, with the noise of small inaccuracies, did the Copernican revolution come about. That is, the "flip" between earth and sun as the singularity did not come about as a result of some force of remarkable logic, or a single monumental piece of evidence, but by small "errors." In the same way, it may be argued that cognitive processes may "switch" around major singular points of thinking by the accumulation of perceptual noise.

The essential stochastic nature of neurobiology redounds to all aspects of what may be called "personality"—from basic neural functions to matters of learning, perception and "cognition." What is remarkable is that these layers of singular stochastic dynamics can most of the time establish a relatively stable configuration. Destruction of portions or all of these pdfs alter the configuration; for example, neurodegenerative diseases alter personality, as can physical trauma. No less is the possibility that a significant perceptual trauma can radically alter the pdfs.

5.3.1 *Examples*

In terms of the clinical psychology of common problems (to differentiate from disorders which may involve psychoses), the well appreciated success of "cognitive" therapies devolve around a patient's ability to overcome historicism's curse: a patient first in some way must be able to accept the fact that past history need not define the present. This often entails the recognition of "alternative" views of the past, which, in effect, decouples that person from the "burden" of the past.

The obvious connection to existential psychology is obvious a propos of the idea that consciousness must be conscious of "something"—past, present—

future (?); i.e., something already a "fait accompli" can "determine" the "being" of a person, and Sartre's idea that man is "condemned to be free." Care, however, must be employed in divorcing the historic past—knowledge of the past does not necessarily imply being determined and connected to the past.

Implicit in these ideas is learning and personality. It is noteworthy that few people have a defined "sense of self" and accompanying memories of a very early age. It may be due to the fact that the "attractor" of personality (as developed by the brain) has not established a defined enough probability of neuronal connections to establish such a distribution: if neuronal connections are essentially uniform in their shape, it is questionable if an attractor is defined. With repetitive learning inputs, the pdfs become established (narrowed) and "personality" emerges. Learning skills proceeds along similar lines: repetitive "habits" further narrow the pdfs so as to make a particular action more refined to the point of not requiring active effort. Both personality and learning, however, are dependent upon the genetics which establish the basic physiology of the neuronal machinery. Predictability regarding personalities and activity is by definition of the singular dynamics, a stochastic process: no matter how narrowed the pdf, there always remains a level of uncertainty.

As a final comment and summation, the dynamics of personal psychology may be summed in terms of what I have already called its "historicism": there seems to be a constant struggle between one's personal history (as developed in one's own stochastic attractor), and the demands of the present and future. In some sense this is the perfect scenario for making life a constant creative struggle along with remarkable possibilities for difficulties. As I have suggested, this seems to be at least part of the program outlined by existentialists.

I would like to suggest, however, this tension is not simply a conceptualization of psychology, but rather follows from the very nature of the construction of the brain: the singular dynamics result in categorizations based upon the stochastic attractor which is the "mind"—"personality." It is an autochthonous, organic result. The challenge is to constantly "update" the attractor in real time—perhaps this is an impossible task while the person is conscious, and sleep is perforce the necessary loss of consciousness required to relieve this tension. Certainly, there have been objective studies demonstrating that learning continues during sleep—and dreams themselves have long been suggestive of the process of integration.

When this integration is not accomplished, is when difficulties arise. Indeed, it is interesting to note that the phrase "let the dead bury the dead" (Matthew 8:18-20) has long enjoyed an interpretation suggesting that a person needs divorce himself from the past in order to allow himself to live effectively in the present.

Such an injunction, however, is not so simple. To totally forget the past would destroy one's notion of "self." As is well documented, certain types of brain strokes (and/or trauma) can result not only in a loss of past memory, but also of the ability of creating new memory. These people are placed in a strange limbo of being conscious but know really knowing who they are—they simply react to the present

People who are strongly tied to their past, however, suffer equally in their repetitive need to deal with their history—their historicism. This is obvious when dealing with traumatic experiences—especially forms of delayed stress disorders. But equally distressing can be positive histories which may prevent effective living in the present.

That some notion of one's past is necessary may in part be supported by the example of so-called "borderline personality disorders," or sociopathy. So-called "moral" or "ethical" existence has been to a certain level documented to be something learned in childhood. Failure to be "connected" to this history may be the etiology of the pathology.

Certainly, the dynamic, "plastic" evolution of the neural system can exhibit numerous difficulties for the "person," inasmuch as the dynamics by definition are not fixed, and are stochastic. But understanding that the dynamics can be in some way stochastically controlled (refer to the discussion on neutron star equations) may provide some realistic solutions.

5.4 *Sociology*

The major way in which singular dynamics can model groups of people is via the understanding that each "discrete" person is in effect a singularity with a "probability distribution" of choices relative to activities related to other people. Thus the compendium of relationships and choices establish a stochastic attractor for the group. Many of these ideas have already been implicit in studies using path analysis as a tool. The only thing lacking is the overall conception of the ensemble as a stochastic process which is characterized by the combinatorics. To some degree, the interest in networks of all kinds (people, animals, and even genes and proteins) perforce exhibit network features. Often a goal is to understand mechanistically, the specific features establishing the connection. In terms of singular dynamics, the fact that these singularities are divorced from deterministic causality emphasizes the stochastic aspects of these connections. It is often pointed out that variability in social science studies is significantly higher than in physical studies. Clearly, this is a result not only of the "person" but also of the person's mental stochastic attractor. This is to say that there are several scales of attractors operative: the "random" sums of squares are frequently large because of the inherent singular dynamics. Moreover, since the attractor is stochastic, it is a mistake to assume that connections (or attractors) remain fixed—they are fluid depending upon the connections. In true nonlinear fashion, a rare event can propel the singular person to a totally unforeseen configuration.

In this respect, its seems questionable to speak of "systems" as a defined entities. The literature has traditional spoken of social "systems" as being unique. Implicit in this is the idea that they combine to work "cooperatively" in some fashion to create organization. It is not clear whether this is necessarily so. Given the stochastic nature, and the independence of the individual "units," suggests instead an addition/subtraction process faintly reminiscent of Bak's "self-organized criticality." The "criticality," however, is really in the eye of the beholder: there is no intrinsic teleological impetus—the dynamics themselves generate the process with the pdf of the individual units establishing local preferences which may or may not redound over the entire network. "Communications" between units often depend upon the "strength" of their connections which are not in themselves deterministic. "Organization" in this context does not seem to be appropriate. We usually think of organizations as groups of individuals with defined goals or purposes. "Natural" organizations have no goals or explicit purposes. They form by accident and their "structure" is totally accidental. There is nothing defined about them—rather, they are evanescent and stochastically driven. Any thing organizational about them is more of the observers need for classification—for forming order. "Emergent structure" exists only in the pdfs of the connections and as such the only real structure is the basic singularity and its attendant stochastics.

5.4.1 *Examples*

Sports such as soccer or basketball are remarkable for the beauty of coordination in excellent teams. The teams appear as analogous to neural networks in that they form a "stochastic attractor," which has been previously discussed: when the players come to know each others "moves," they can effortlessly pass the ball from one to another with hardly a glance, and with a high degree of confidence in that the subsequent players will move the ball in the direction desired. This is, in effect, a "narrowing" of the probability distribution relating the players' action in a given circumstance, and is effected by extended practice and learning. There is no direct control between the "neurons" (players) except through their individual knowledge that each other player will "probably" act in an appropriate way.

A similar scenario may be found in a well-run corporation: the individual managers may learn to "play" together, such that communication is minimized (they already know what they will do). By reducing unnecessary communications, the company is more efficient, and can react to changing market circumstances quickly and competitively as compared to their less coordinated competitors. It also points out the dangers involved in the elimination of what might be perceived as "unnecessary" managers. Although, strictly speaking a given manager might not be crucial to the operation of an organization, that manager may have knowledge of interconnections which make the organization function in a highly "productive" manner. Elimination of such a manager will not cause the demise of an organization, however, it will work less efficiently until, some other manager "learns" the connections.

Frequently, organizations relate many of these concepts to the idea that employee "turn-over" and the attendant costs relate to the need to "train" new employees. It is questionable whether an employee can absorb a significant amount of information in a short period of time. What perhaps really occurs is a longer period during which the employee comes to learn the appropriate "probabilities" associated with other employees.

Finally, there is the question again of how can these social aggregates be in some way manipulated for the over-all benefit? It would seem that the paradigm of injecting "change" at certain critical junctures of the nondeterministic process, can increase the probability of a desired effect. Certainly this has been well know in a tacit sense as judged by examples whereby certain small pieces of information injected into a social network are designed to effect a certain social change. There is no guarantee that the desired effect will develop, however. Market advertising specialists have known this for a long time: their advertisements are often subtle—strong suggestions are less predictable or successful. Thus it is often the subtle context of a presentation that makes the most effective change.

Clearly, the ethical context of such manipulations is more complicated when one is dealing not only with the group's "history," but also the individual histories of the people involved. However, as a limiting factor, everything redounds to the initiator's ethics.

6 *Conclusions*

The objective of this monograph was to suggest a different paradigm for viewing dynamical science. The paradigm is based on discontinuous dynamics which are strongly conditioned by singular points that are sensitive to the effects of noise. The motivation for this model was an attempt to explain reality as it is: there is much determinism, but there is also much randomness, and there seems to be an interplay between the two in a way by which the two inform each other. In a universe which appears to be composed of vast amounts of "noise," it would seem to be improbable that this noise would not be somehow involved in the dynamics in some important, relevant way. Moreover, this view, contrary to a first impression, is simple—it follows Occam's razor: by postulating singularity-based dynamics, the dynamics are more easily understood by virtue of their flexibility to model a variety of probabilistically-based motions where they actually occur, as opposed to assuming a complicated, multi-variable deterministic process. Additionally, the techniques to identify singularities, etc., are relatively simple, using at their basis recurrences.

Certainly, many biological or physiological variables exhibit an apparently random behavior which is often punctuated with seemingly deterministic processes. Tests of experimental time series of physiological signals, demonstrate some basic problems associated with attempts at characterizing their real deterministic-chaotic nature. This is to emphasize the problem of establishing the actual features of the assumed existing determinism in the dynamics of living systems.

It is not the aim here to completely explore this problem, but rather to adduce some new results giving a possible support to the thesis that there is a class of phenomena that cannot be represented by chaos directly, but rather by discrete events dynamics where randomness appears as point events: there is a sequence of random occurrences at fixed or random times, but there is no additional component of uncertainty between these times. One of the main features of living matter is its continuous adaptive requirement throughout a range of time scales and the discrete events dynamics is expected to have a basic role in realizing the basic control mechanisms of the whole biological dynamics.

As has been suggested, an implied determinism represents the basic paradigm of science. It admits the proposition that each event is necessarily and uniquely a consequence of past (future) events. This implies that a major feature of determinism is its rigidity. Deterministic systems exhibit a strong dependence of the system from initial conditions in the sense that at any time evolution is regulated from strong links to the initial conditions from which they started. Chaotic determinism exhibits the same rigidity since, in spite of the chaotic behavior of the system, the strong requirement of the acting deterministic rule remains. In living matter it is difficult to accept such an extreme dependence of a physiological system on its initial conditions. A physiological system, in fact, is required to adapt continuously its behavior to the requirements imposed from the environmental conditions in which it operates.

Moreover, biological systems often exhibit oscillations in their behavior which are not strictly periodic; instead, the signals exhibit "pauses," which cannot be indicative of deterministic behavior. They indicate nonstationarity and moreover the question of their correct interpretation arises as the basic problem that one must solve in order to coherently describe the dynamics.

In mathematical determinism, the basic key is represented by Lipschitz conditions whose validity ensure uniqueness of solutions for differential equations and thus uniqueness of trajectory in the dynamics of the corresponding system. The previously mentioned pauses observed in various data, may be considered to be an expression of singularities arising in the mathematical counterpart describing the physiological phenomenon. We are thus concerned with the particular implication of non-Lipschitz equations, since these equations admit non-unique solutions. If a dynamical system is non-Lipschitz at a singular point, it is possible that several solutions will intersect at this point. Since the singularity is a common point among many trajectories, the dynamics of the system, after the singular point is intersected, is not in any way determined by the dynamics before, and thus, in conclusion, this system will exhibit a dynamics that may no longer be considered deterministic. The system will have a nondeterministic behavior and an immediate consequence of such nondeterministic dynamics will be the possibility of nondeterministic chaos.

The support for the traditional mathematical approach to the description of these phenomena has been the formal, classical logic of Aristotle. Although this approach has clearly produced remarkable results in the history of Western civilization, developments in the 20th century have served to question the exhalted position of this logic. Certainly with the development of quantum physics, probability has come to be more appreciated beyond its interest in the science of games. And "noise" has come to be appreciated as more than a nuisance for determinism. What may be important is the recognition that noise is not outside dynamics but is an integral part of and generated by the dynamics.

Given that I have tried to promote the importance of probability and nondeterminism, I need to emphasize that "determinism" persists in most of the models presented. The difference is that it is "piecewise." In some physical systems, and depending upon scale, the determinism dominates. In other systems, the determinism is still there, but perhaps a bit more diminished.

And finally, nondeterminism should be considered an ontological entity—not dependent upon epistemological imprecision.

7 *Glossary*

In order to help non-specialists, these brief definitions of some important terms are presented. Again, as with the general tone of this work, the aim was to present basic concepts as opposed to scientific rigor.

Attractors. A term used to refer to the long time behavior of a dynamical system. Roughly speaking, the attractor draws all points (states) to it. Another way to look at it, is to say that it is emblematic of a kind of stability. Attractors can be further classified as, e.g., fixed points, limit cycles

Bifurcation. As the Latin derivation implies, the dynamics qualitatively change due to a change of parameters; e.g., from period one to period two.

Chaos (deterministic). In terms of the popular idea of "chaos theory," the appellation is somewhat of a misnomer. The popular notion of the word suggests disorder, or confusion. The actual theory, however, is based on the concept that fully deterministic (i.e., rule-obeying) equations can produce an output which appears totally chaotic (i.e., not rule-obeying). Initially, this motivated many scientists to consider the possibility that many natural systems are similar: they look disordered, but they, in effect, obey specific rules. Unfortunately, this does not seem to be the case. The original idea was based on low dimensions, whereas many real systems are high dimensional. Furthermore, there are many alternative ways of explaining apparently chaotic dynamics. Nevertheless, this lead many investigators to look for the possibility of chaos by calculating measures which could suggest chaotic dynamics. Regrettably, many of the methods for calculating these invariants have specific mathematical assumptions which are rarely met by real systems. Consequently, an uninformed use of these methods often results in equivocal results. There is no "official" definition for chaos, but there are several technical requirements, which are often rarely addressed. All in all, there are many aspects related to the description of a system before it can be suggested to be "chaotic." This has not, however, stopped many from declaring that various systems are chaotic, which has lead to much confusion. This state of affairs led a respected scientist in this area, David Ruelle, to question the objectives of simply naming a dynamic as "chaotic"

Complexity. Although this term is often used both in common language as well as science, there is no adequate description. Indeed, every year presents new formal definitions. Many of the definitions depend upon the perspective of the viewer. Some people stress the underlying dynamics which generate the observed "complexity," whereas others focus on the output itself. Thus the same dynamics may be considered complex by one set of judges, but not by another.

Degree of Freedom. Means different things in different contexts and/or disciplines. Physicists often use the term with respect to Hamiltonians, and their conjugate pairs, p, q, (see below). Other times it is used more generally, with less precision—i.e., the minimum numbers of parameters needed to describe a system. Statisticians, on the other hand, usually refer to the number of independent random variables which constitute a given statistic—often this means one less than the number of variables.

Determinism. A term which has much philosophical connotation. Strictly speaking, a system is deterministic if it is rule-obeying, and consequently its evolution can be predicted. In "chaos theory" these systems are, in fact, deterministic, since they obey rules, and their evolution could be predicted if they were known with infinite precision. Of course, infinite precision is not possible, and consequently their future cannot be easily predicted. The idea that natural phenomena could be described by such rules probably gained impetus with Newton's laws. His and his heirs' relative success spurred efforts to find additional "laws" for a "mechanistic" world. Unfortunately, especially with the advent of quantum mechanics, it became clear that such a mechanistic world view could not be supported. In fact, many of these laws worked because of their approximation to reality with respect to a given scale of measurement. It was Einstein who pointed out that mathematical models are only an approximation of reality. The mathematics are often an idealization of the phenomenon (which is not to disparage their usefulness). Thus it is important to realize the separate spheres of mathematics and reality. There has been an ongoing debate regarding reality relative to the deterministic/random dichotomy. The question has always been one of determining if "things" are ordered or happenstance. One can trace these ideas from Aristotle on to contemporary science. Perhaps a more important question is the agency of the brain in such perceptions. Modern scientists have demonstrated that the brain is "pattern-seeking" –and finding—even if no real pattern exists. Clearly, the evolutionary benefit (based upon a personal history) is clear. The other consideration is what has been argued here; namely, that patterns or determinism need not be continuous. It may be that randomness is the usual case and that determinism punctuates local spaces. Perhaps so much debate regarding the term would exist, were it not for its counterpole; namely, freedom or "free will." Inexorably, the two terms are often brought together especially when it involves humanity vis-à-vis scientific inquiry. These terms seem always to be mutually exclusionary. The suggestion here is the nondeterministic dynamics can accommodate both without contradiction.

ECG. The electrocardiogram—a record of the electrical signals of the heart recorded from the skin surface. Because it is a surface signal, it is in some sense an amalgamation of all the different unique voltages of the heart. Besides being correlated with certain cardiac and pathologic events, it does not provide significant detail regarding localized cardiac activity. For this reason, more invasive techniques are used to pinpoint areas of particular electrical activity.

EEG. The electroencephalogram—like the ECG is a surface recording, based from scalp electrodes, of the millions of microvoltages discharged by brain neurons. Again, this is an amalgamation of all these signals. For precise localization, again invasive recording techniques must be used.

Fast Fourier Transform (FFT). A technique which reduces the time necessary to perform a Fourier analysis on a computer.

Hamiltonian. A function H, used by physicists as a rewriting of Newton's laws, such that:

$$\frac{\partial p_i}{\partial t} = \frac{\partial H}{\partial q_i}; \frac{\partial q_i}{\partial t} = \frac{\partial H}{\partial p_i}. \qquad (7.1)$$

where p_i and q_i are state space variables—the whole equation relating to energy. Such a system is termed conservative is these state-space variables are asymptotically invariant, dissipative otherwise.

Fourier Analysis. Decomposes a signal (time series) into a finite sum of trigonometric components—sine and cosine waves of different frequencies. Named after the French mathematician of the early 19th century. The mathematical assumptions are that the data are stationary and linear. If the data are short or transient, or nonlinear, the results become very difficult to interpret. Small changes in the series are especially difficult to evaluate given that they are often "smoothed" out by the dominant periodicities in the process. Many "workarounds" have been devised including partitioning the data into smaller windows, as well as modifications of the mathematics such as the Wigner-Ville transform or "adaptive" methods with varying levels of success.

History. The idea that history should be included in a specialized glossary, at first may seem strange. However, it is placed in the context of human brain perception. It can be argued that what constitutes a person is the person's history as developed by the vast pattern-forming apparatus of the brain. As is well known, the history—the memories are not invariant—they can be changed by a variety of factors, and their influence on "personality" and motivation are in turn influenced by the physiology of the brain and vice versa as increasingly modern neuroscience has demonstrated.

Instability (stability). There are several formal mathematical definitions, but, in general, one can say a point is stable if it tends to stay in the same neighborhood of the dynamics, but unstable, if it tends to fly off. Things can become more complicated in terms of distinctions applied. For example, one can speak of local, asymptotic, global or structural forms. One measure commonly used in this respect are Liapunov exponents (or Lyapunov, depending upon the transliteration scheme from the Russian Cyrillic alphabet), named after the early 20th century Russian mathematician. If the exponents are positive, the system is chaotic—or at least diverging. There are several other methods depending upon the type of dynamics.

Invariant. A technical mathematical term which defines quantities that distinguish the relevant properties to be included in a given class. This is to say these are characteristic, distinctive features.

Langevin Equation. A type of equation used to model a particle's behavior with stochastic forcing. There are many such equations for specific assumptions.

Lagrangian. In mechanics, it is the difference between the kinetic energy and the potential energy of a set of particles, and is expressed as part of an equation using generalized coordinates.

Laplace. Famous 18th-19th century physicist and mathematician who is often quoted in connection with the classical scientific view of determinism. He said: "We ought to regard the present state of the universe as the effect of its antecedent state and as the cause of the state that is to follow. An intelligence knowing all the forces acting in nature at a given instant, as well as the momentary positions of all things in the universe, would be able to comprehend in one single formula the motions of the largest bodies as well as the lightest atoms in the world, provided that its intellect were sufficiently powerful to subject all data to analysis; to it nothing would be uncertain, the future as well as the past would be present to its eyes. The perfection that the human mind has been able to give to astronomy affords but a feeble outline of such an intelligence." Pierre Simon Marquis de Laplace, *A Philosophical Essay on Probabilities*, 6th edition, translated by F.W. Truscott and F.K. Emory (New York: Dover, 1961), p. 4. Excepting the 19th century world view, as many have noted, there is little difference in this exposition from Aristotle's reflections in his *Physics* on causality. The difference perhaps lies in how the causality is determined. As the famous statistician noted, correlation expresses the same phenomenon without necessarily resorting to determinism.

Linear. Referred to dynamics which exhibit the principle of superposition. This is to say that system inputs produce proportional outputs. In its most elementary presentation,

$$f(x+y) = f(x)+f(y) \text{ and } f(ax) = af(x). \tag{7.2}$$

Neural Nets. Are models algorithms for cognitive tasks such as learning, and are loosely based on an imitation of the brain. Mathematically, they are considered directed graphs. Each "neuron" or node is a state variable linked with others in terms of a "weight" and "transfer function." Obviously, the idea is model each neuron through synapses (links) which follow the transfer function rules. Early models aimed primarily at developing information processing algorithms. More recently, they are increasingly attempting to mimic actual brain processing. The models can become quite complicated relative to their structure and "rules."

Noise. A difficult term to define. Commonly this refers to unwanted, disordered sound. Visually, it is the "snow" sometimes seen on a television set. To a scientist it may mean random, featureless data. It usually connotes a random signal, and the opposite of determinism. In practice, it may be the result of all those factors which cannot be adequately explained, or accounted for. Often it is described as having "colors," such as "white" (uniform spectrum), or "brown" (high frequencies damped out). It can also be described in terms of it power law "scaling;" i.e., white scales with respect to frequency (on log log scales) as f^0, brown as f^2, or a form typically found in nature as f^1.

Nonlinear. Usually a system is defined to be nonlinear by its negative; i.e., it is not linear. This bit of circular reasoning is obviously problematic. The reason is that nonlinear systems can be so complex as to defy an easy explanation. But basically, such a system does not have simple proportional relationships between input and output. Often, this implies that small input changes can produce a relatively large and perhaps unexpected effect.

Non-Lipschitz Dynamics. Dynamics not characterized by a mathematical formalism which demonstrates that a function is continuously differentiable. This is to say the function may not be "continuous, and instead have "jumps." This defines mathematically one form of nondeterministic dynamics.

Nondeterministic Dynamics. By our definition, dynamics characterized by singularities, at which points the past is disconnected from the future, and is governed instead by probabilities. At times, authors use "nondeterministic" dynamics to refer to random processes. As can be seen, in our definition, this is not quite the same, since although there is randomness, there are also deterministic dynamics intermingled between the singular points.

Norm. Roughly, a way to measure the length (magnitude) of a vector in n-space. There are several common ways to do this; namely, the min norm, Euclidean norm, and max (or supremum, infinity) norm. Each has certain qualities, such that, the min norm emphasizes details; the max, the broad outline; and the Euclidean, a compromise between the two. Theoretically, any can be used, but often they are chosen on the basis of computational effort: clearly, the Euclidean, requiring squaring and square root operations, takes longer to compute. For theoretical reasons, they are often compared in terms of the "unit circle." The concept of unit circle (the set of all vectors of norm 1) is different in different norms: for the min norm the unit circle in two dimensional space is a rhomboid, for the Euclidean norm, it is the unit circle, while for the max norm it is a square (Fig. 7.1).

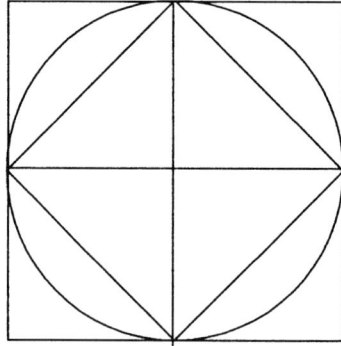

Figure 7.1. "Unit circle" of min norm (rhombus, innermost); circle of Euclidean; and square of max norm.

Patterns. In some sense, the objective of science is to discover patterns, and why they are there. To some extent, this is related to the notion of determinism; i.e., there is some "rule" guiding the development of the pattern. Implied is the idea that patterns create meaning, etc. The argument here has been that pattern formation is a natural function of the brain—even if none is there.

Phase Space. A space of $2s$ dimensions representing a system with s degrees of freedom. It is a union of coordinate and momentum spaces in which the rectangular coordinates represent the position and momentum of the points of a system. By plotting the points, the time evolution of the system can be followed (Fig. 7.2). In practice, however, all the necessary information cannot be obtained. This resulted in the so-called "state space reconstruction" using delay coordinates of a single variable, whereby:

$$\begin{aligned} x_1(t) &= v(t), \\ x_2(t) &= v(t-\tau), \\ &\vdots \\ x_d(t) &= v(t-(d-1)\tau), \end{aligned} \quad (7.2)$$

where τ is a delay time and d is the dimension.

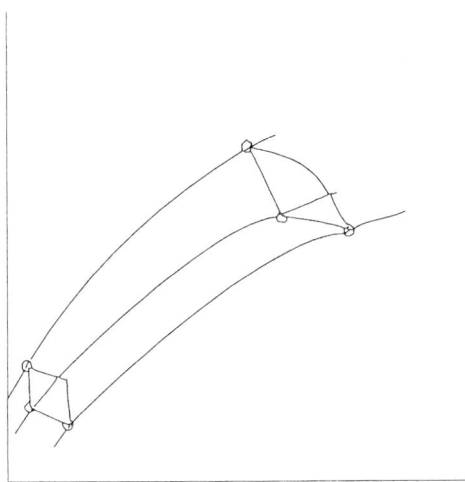

Figure 7.2. Evolution of a volume in phase space.

Power Spectrum. Is usually a graphical representation of the results of spectral analysis. Usually expressed as the square of the Fourier transform. The term "power," according to some sources reflects its frequent usage for calculation of energy/power relationships. It is important to stress that, when a signal is described by its power spectrum it is projected into a completely different frame of reference with respect to its original space. In fact we are used to embed time series in a bivariate space in which the abscissa represents time and the ordinate the varying values of some measure of interest. This kind of graphical representation allows us to locate in time each particular noticeable feature of our measure, so we can detect when the system changes rapidly and when it stops, and its relative increases and decreases. When looking at the power spectrum of the same signal, while we acquire the general view of the periodicities (i.e. regularities) present in the signal, we loose any possibility to locate them in time, the independent variable of power spectra is no longer time but frequency—the description corresponds to a general average over all the signal. As a result, although periodicities can be defined, their "location" in time is not indicated. Power spectra shift the observation lens to the "good velocity / bad location" side, this implies that if the system is stationary, power spectra work in an optimal way because the location information is not relevant given the system will come back indefinitely on the same "places" and knowing what frequencies are present, any possible location is a sufficient information. On the contrary, for an extremely non-stationary system the information of the power spectra is relatively poor. This trade-off between location and velocity (energy) has been noted by generations of signal analysts who have devised may kinds of mixed strategies for arriving at a compromise optimal for the investigated systems. This includes wavelet analysis, moving FFT, and maximum entropy methods. The recurrence histogram is again a sort of power spectrum and, like power spectrum, loses the explicit location information but, given the local character of recurrence estimation that is based on pairwise independent comparison between short epochs, it does not impose (at odds with power spectra based on Fourier Analysis) a general explanation model over all the signal.

Probability Density Function (pdf). A statistical term with reference to the function representing the frequency of a continuous random variable from which such parameters as mean and variance can be derived. Practically it can be suggested from the histogram of data—as the amount of data increases and the intervals decrease. Two or more random variables are described by their joint density function. Sometimes referred to as simply the density function.

Proteins. An important class of molecules necessary for life, which are composed of amino acids. Each protein found in nature has a specific three-dimensional structure and this structure is determined by the sequence of amino acids. This makes the particular linear arrangement of amino acids constituting a protein an efficient recipe for the solution of a chemico-physical problem that basically is the folding to a given unique three-dimensional structure in water solution. How this is achieved has been an on-going topic of investigation. Proteins are extremely important in carrying on the "work" of living matter.

Randomness. Mathematically, something is said to be random if its value can only be described probabilistically prior to actually taking the value. However, more generally, something is said to be random if its causes cannot be discerned or appears to be devoid of any structure. Correlations are often used to determine how "random" a series is: presumably the less the long term correlation, the more "random" it is. This, however, is somewhat of an artifice insofar as real "random" systems do admit correlations. Consequently, there is some disagreement as to what constitutes a truly random series.

Recurrence. (and recurrence plots and quantification). See Mathematical Appendix.

Scaling. Relations of the form

$$f(x) = k\, x^h \qquad (7.3)$$

are called power law relations. If a system is self-similar, i.e., there is some feature that is constant on all scales, it can be represented by something called a power law or scaling law. Power laws are a simple description of how a system's features change in proportion to the scale of the system. When viewed on a log-log plot the relationship is linear. Scaling laws can be viewed as a simple way with which to describe aspects of complex systems. Such laws are insights into the fundamental processes and constraints that operate on or within a system. The slope of a scaling law is thus a precise and repeatable reflection of the rules that operate in a system. Systems that are described by power laws identical in slope suggest similarities in the underlying rules or processes; differing slopes point toward basic differences. Care, however, must be used in determining if the relationships are real. An understanding of the system is important. Various processes can generate power laws, so that identifying something as having a power law is essentially a first step in its analysis.

Singularity. There are many definitions depending on the discipline. Usually defined by mathematicians as a point which is not differentiable such as a curve which does not have a smooth tangent. A more intuitive way to view this is as a discontinuity. Physicists tend to look at singularities as quantities which are infinite, or where the usual "laws" break down. (Stephen Hawking has made this idea very popular in his writings on cosmology.) Sometimes they speak of a place where things "blow up." Thus the usual theory governing a situation may not apply. Some of Eisnstein's ideas express singularities. Very often, especially with respect to mathematics, a singularity is to be avoided, and a variety of manipulations can be devised to transform a singularity into something more mathematically tractable.

Spectral Analysis. General term for a number of mathematical techniques used to describe the frequency content of a signal. Typically, the results are plotted in graphical form, with the y-axis indicating the "amount" of a given frequency (x-axis). Fourier analysis is one form. At times the term is reported without specifying the particular methods. For example, there are parametric and nonparametric methods, which can produce contradictory results depending upon the specifics. For this reason, authorities sometimes point out that this analysis is often as much art as it is science.

Stationarity. A common assumption in many time series techniques is that the data are stationary. Unfortunately, many scientific papers present results based upon techniques which require stationarity, but no evidence is given. A dynamic is said to be stationary if the mean and variance do not change over time. How to determine this can be problematic, since simple inspection of a time series that is windowed is frequently equivocal (e.g., moving average), since the width of the window determines the degree of smoothness (stationarity) obtained This problem is especially difficult if one is dealing with a nonlinear system, since most tests are not made for nonlinear systems. This is why a technique such as recurrence analysis can be useful, since it does not require an assumption of linearity. Alternatively, the series can be searched for locally stationary areas. Still other authorities claim that stationarity can be obtained by applying a transformation such as "differencing" the series.

Stochastic (Process). Any process that can be described by a random variable usually in time. When indexed, they describe states related to probability.

Systems. Especially after WWII biology has been strongly influenced by an approach loosely termed "Systems Theory." There are many "flavors" but include contributions from von Bertalanffy's systems theory, Wiener's cybernetics, Shannon's information theory, and von Neumann's game theory. Some take it further back to the idea of Laplace's and Newton's mechanistic determinism. Because of its diverse background, it is dangerous to attribute specifics, but there is a tendency to emphasize interrelatedness of elements, and as a result, the possibility to model the overall organism. As has been suggested, however, this may suggest a logical organicism which is not always apparent. One difficulty with this approach is its lack of understanding "noise." Concomitantly, there is a strong "engineering" flavor to the discourse, with common usage of terms such as "control," and "set points." A possible result has been the emergence of the ideas of "chaos" and "complexity." Certainly the difficulties of an excessive faith in such approaches have resulted in a gradual recognition that modeling from such a perspective is not always facile. Consider, for example, the difficulty of explaining phenotype from genotype: very small differences between species genes effect a remarkable difference in the final "product." And although interrelatedness is certainly an important feature of organisms, the links, as has been argued, are characterized by singularities—not continuous processes. Consequently, the insights gained should not be ignored but modified. Thus, as has been recommended, an optimal metaphor is not a machine, but a map of a city, with its accretions, deletions, connections, and autonomous sections—all modifying their links in response to changing circumstances. More recently, the systems approach has evolved into views of "self-organized criticality"—a view that proposes that systems organize themselves into near critical states. A commonly used example is that of a sand pile, which with the addition of grains can suddenly result in avalanches, which further result into a new state to "relieve" the critical state. Although it appears that many phenomena behave in such a manner (and may exhibit power laws), it is not always clear how attendant circumstances modify this paradigm. This, in effect, concerns the relevant "boundaries" of the involved observable.

Transient. A term more often used in experiments or observations as opposed to mathematics. The idea is that the observable is, with respect to its state, temporary, and not really characteristic of a particular system. As such, if it can be identified, it should be ignored. Mathematically, it is what happens before the variable settles down to its "attractor." The difficulty, of course, is in identifying a series as a transient. There are no good guidelines to help an observer determine if a series is transient. Realistically, it comes down to relative observations, and the existence of some prior or post "stability."

Uniqueness (and Existence) Theorems. Here referred to with respect to the theory of differential equations, which requires that their solutions not only exist, but are also unique. This is usually given as a proof based on the idea that there are two different entities that satisfy given conditions, but then demonstrating that this is not possible. There are several such theorems, and the

"Lipschitz conditions" are one of these. Depending upon the particular theorem, the proof may be defined only for certain boundary conditions.

8 Mathematical Appendix

8.1 Nondeterministic System with Singularities

It has been pointed out that the traditional deterministic Laplacian view of dynamics, which assumes that the time evolution of a system is uniquely determined by its initial conditions and past behavior, may be at odds with some real systems. The traditional tenet devolves from the "existence and uniqueness theorem" of differential equations; i.e., they are Lipschitz continuous. Yet, there is nothing in classical mechanics that requires Lipschitz continuity. In fact, there is no a priori reason to believe that the Laplacian model holds true in every case. For lack of a better term I have variously called such dynamics "piece-wise deterministic," "terminal," and "nondeterministic."

I have argued that there exist many physical and biological "motions" which may be better modeled by non-Lipschitz differential equations whereby there is no single solution to the equation. Instead, the problem becomes a combinatorial one with a resultant "nondeterministic" chaos, and "stochastic attractor." This is to say that each singularity has an associated probability distribution regulated by factors unique to the given system. Based upon this phenomenon as a paradigm, a dynamical system whose solutions are stochastic processes with a prescribed joint probability density can be developed. The general form of such differential equations is:

$$\dot{x} = -x^{1/3} \qquad (8.1)$$

$$\frac{d\dot{x}}{dx} = -1/3x^{-2/3} \to -\infty, x \to 0. \qquad (8.2)$$

This equation has an equilibrium point at $x = 0$ at which the Lipschitz condition is violated.

The relaxation time for a solution with the initial condition $x = x_0 < 0$ to this attractor is finite:

$$t_0 = \int_{x_0}^{N \to 0} \frac{dx}{x^{1/3}} = 3/2 x_0^{2/3} < \infty. \qquad (8.3)$$

This represents a singular solution which is intersected by all the attracted transients. For the equation:

$$\dot{x} = x^{1/3}, \qquad (8.4)$$

$$\frac{d\dot{x}}{dx} \to 1/3x^{-2/3} \to -\infty, x \to 0, \qquad (8.5)$$

the equilibrium point $x = 0$ becomes a repeller: If the initial condition is infinitely close to this repeller, the transient solution will escape the repeller during a finite time period:

$$t_0 = \int_{\varepsilon \to 0}^{N \to 0} \frac{dx}{x^{1/3}} = 3/2 x_0^{2/3} < \infty, \qquad (8.6)$$

if $x < \infty$,

while for a regular repeller, the time would be infinite. Indeed, any prescribed distribution can be implemented by using non-Lipschitz dynamics.

Thus, it is important to detect such singularities so that apparent complexity is not mistaken for actual complexity in the underlying system. An important feature of such systems is the abrupt change of the trajectory as it changes from one solution to another as it approaches the singularity. This jump is essentially instantaneous, and the solution is

$$x(t) = \Theta(t-t_0)x_1t + \Theta(t_0-t)x_2t, \qquad (8.7)$$

where $\Theta(x)$ is the unit step function, and t_0 is the time the jump occurs. If $x_1(t)$ and $x_2(t)$ are non trivial and different solutions, then some time derivative of $x(t)$ will diverge at t_0. Typically, the divergence occurs at the second (or higher) time derivative. The ramifications for such dynamics may be especially important for modeling biological processes such as the nervous system, where "natural singularities" in the form of synapses exist.

Almost all experimental data currently is obtained through digitization to accommodate computer storage and manipulation. As is well known, the Nyquist theorem governing the sampling rate is based on linear systems theory; i.e., in order to recover all Fourier components of a periodic waveform, it is necessary to sample more than twice as fast as the highest waveform frequency v, or $f_{Nyquist} = 2v$. The motivation certainly, is not only that the signals of interest are captured, but also that it be done efficiently; i.e., to prevent oversampling with its increased storage burden.

It should be noted, however, that signals which are discontinuous, nonlinear or nonstationary are not addressed by the Nuquist theorem. Additionally, real-life signals are usually prefiltered before the A/D conversion stage to avoid aliasing distortion as well as to limit unwanted signal components. Clearly, if a singularity were sampled, it is conceivable that it would be missed given the overall sampling requirements, i.e., the sampling could be insufficient to locate it; or, prefiltering could easily distort it. Furthermore, the inherent noise of sampling and the system make the second derivative practically useless.

8.2 Recurrence Quantification Analysis (RQA)

The notion of a recurrence is simple: for any ordered series (time or spatial), a recurrence is simply a point which repeats itself. In this respect, the statistical literature points out that recurrences are the most basic of relations, and it is important to reiterate the fact that calculation of recurrences, unlike other methods such as Fourier, Wigner-Ville or wavelets, requires no transformation of the data, and can be used for both linear and nonlinear systems. Because recurrences are simply tallies, they make no mathematical assumptions. Given a reference point, $\mathbf{X}_0$, and a ball (B) of radius r, a point is said to recur if

$$B_r(\mathbf{X}_0) = \{\mathbf{X}: ||\mathbf{X} - \mathbf{X}_0|| \le r\}. \qquad (8.8)$$

A trajectory of size N (i.e., samples) falling within $B_r(\mathbf{X}_0)$ is denoted as

$$S_I = \{\mathbf{X}_{t1}, \mathbf{X}_{t2},...,\mathbf{X}_{ti},...\} \qquad (8.9)$$

with the recurrence times defined as

$$T_1(i) = t_{i+1} - t_i; \quad i=1,2,...,N. \qquad (8.10)$$

If the dynamics is stationary, the reference point can be chosen arbitrarily. In most biological contexts, this cannot be assumed, yet, this very condition is revealed by contextual changes seen in a plot of recurrences.

8.3 Recurrence Plots

Given a scalar time series $\{x(i) = 1, 2, 3, ...\}$ an embedding procedure will form a vector, $\mathbf{X}_i = (x(i), x(i + L), ..., x(i + (m-1) L))$ with m the embedding dimension and L the lag. $\{\mathbf{X}_i = 1, 2, 3, ..., N\}$ then represents the multi dimensional process of the time series as a trajectory in m-dimensional space. Recurrence plots are symmetrical $N \times N$ arrays in which a point is placed at (i, j) whenever a point $\mathbf{X}_i$ on the trajectory is close to another point $\mathbf{X}_j$. The closeness between $\mathbf{X}_i$ and $\mathbf{X}_j$ is expressed by calculating the Euclidian distance between these two normed vectors, i.e., by subtracting one from the other: $||\mathbf{X}_i - \mathbf{X}_j|| \le r$ where r is a fixed radius. If the distance falls within this radius, the two vectors are considered to be recurrent, and graphically this can be indicated by a dot.

An important feature of such matrixes is the existence of short line segments parallel to the main diagonal, which correspond to sequences (i, j), $(i+1, j+1)$, ... , $(i+k, j+k)$ such that the piece of $\mathbf{X}_j$, $\mathbf{X}_{j+1}$, ..., $\mathbf{X}_{j+k}$, is close to $\mathbf{X}_i$, $\mathbf{X}_{i+1}$, ..., $\mathbf{X}_{i+k}$ in series which are deterministic. The absence of such patterns suggests randomness (Figs. 8.1-8.3).

Thus recurrence plots simply correspond to the distance matrix between the different epochs (rows of the embedding matrix) filtered, by the action of the radius, to a binary 0/1 matrix having a 1 (dot) for distances falling below the radius and a 0 for distances greater than radius. Distance matrices are demonstrated to convey all the relevant information necessary for the global reconstruction of a given system, and thus represent an exhaustive representation of the studied phenomenon.

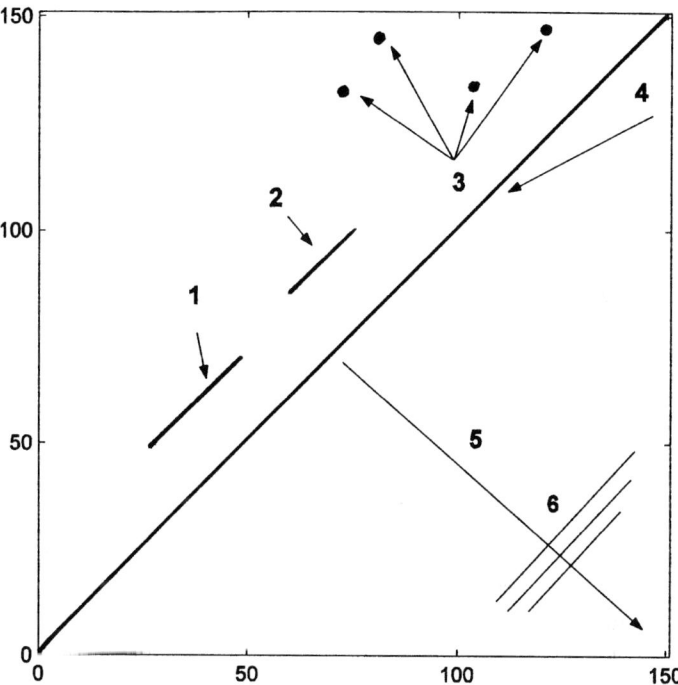

Figure 8.1. Basic features of recurrence plots: 1-2) deterministic line segments—at least 2 consecutive points—these may be redefined (if these deterministic line segments achieve a box-like configuration, then the vertical distances can be computed to determine a feature called "laminarity"—very often these laminar areas can be indicative of singularities, however care must be exercised to determine if there is a shift in the trajectories; 3) isolated recurrence points, which may or may not be random; 4) identity line; 5) direction of time (slope) which is computed over recurrent points (because of poor statistics as the arrow approaches the corner, the lest three rows are not counted in the computation of the trend—this too may be redefined 6); and may be used to indicated trend. Counting the points between diagonal structures can give the period of an oscillation provided the sampling rate is known.

186   *Unstable Singularities*

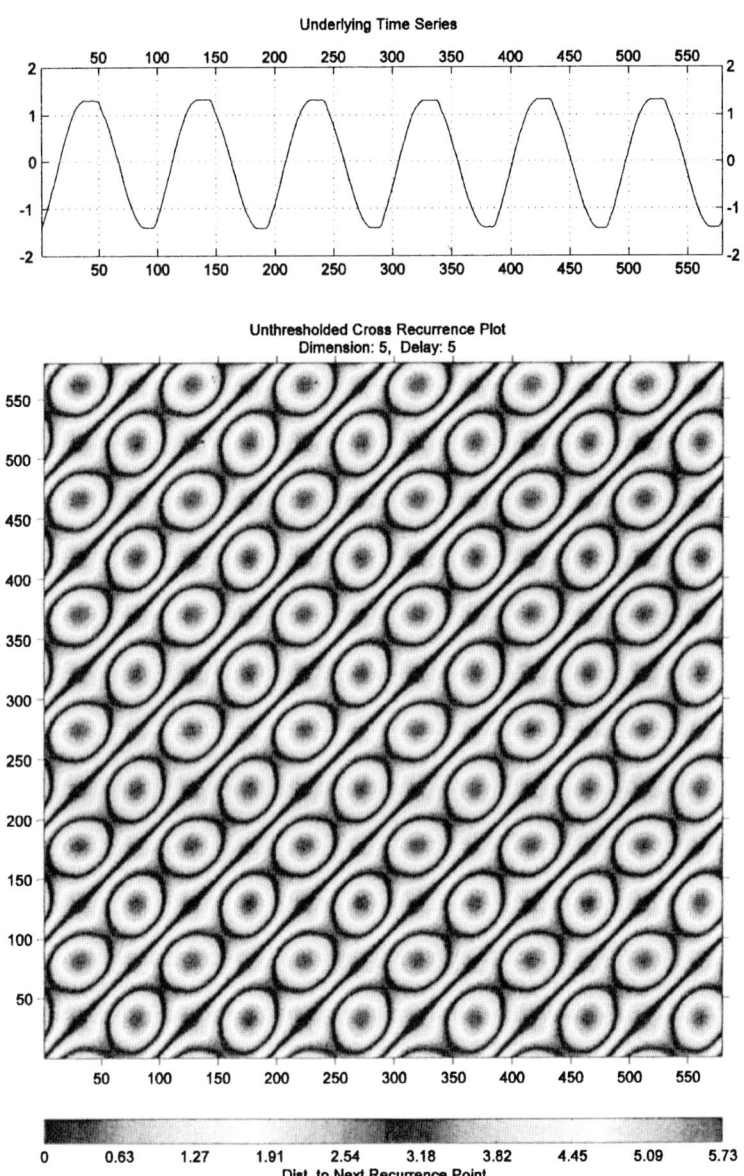

Figure 8.2. Recurrence plot of a sin-wave process courtesy of Norbert Marwan of the University of Potsdam. Here the recurrences are not thresholded, and instead are coded by shades of gray.

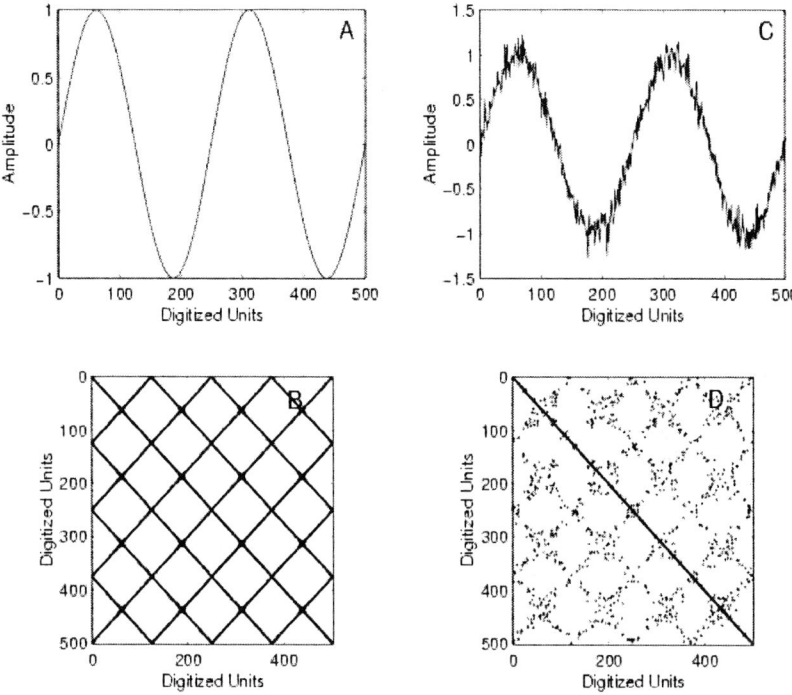

Figure 8.3. Sine waves without (A) and with (C) noise and their representation by recurrence plots (B, D).

The data obtained can also be used to obtain estimations of local Liapunov exponents, information entropy, or simply plotted as N_{Rec} vs. period; i.e., a histogram of recurrence times. In the case of histograms, strictly periodic points demonstrate instrumentally sharp peaks; whereas chaotic or nonlinear systems reveal more or less wider peaks depending upon the radius chosen and noise effects (Figs. 8.4, 8.5). The histogram of the distribution of recurrence times gives a sort of "nonlinear" analogue of a Fourier power spectrum and can be profitably used to derive scaling profiles of the studied systems. RQA can also be combined with other statistical techniques. A particularly useful technique is that of principal components analysis (singular value decomposition), which can "combine" the results of the multiple RQA variables into one.

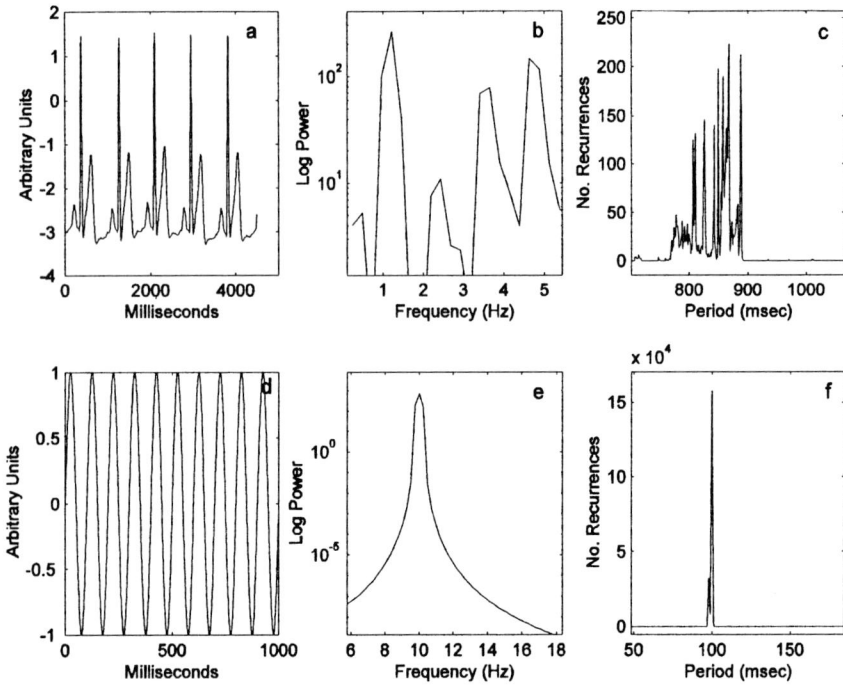

Figure 8.4. Comparison of ECG (a) and sine wave (d) analyzed by FFT (b,e) and recurrence histograms. Note the sharper peaks and greater detail with recurrence histograms. For complicated series, there is less chance of combining nearby frequencies into one.

Principal components does this by establishing orthogonal vectors which independently account for a significant amount of the variance explained by the studied variables. This is done successively until the variance is exhausted. In many cases, these "components" can be uniquely identified or characterized with respect to features of the observed series. Thus it is possible to develop components which signify "size," or "shape." More importantly, components which may not explain significant amounts of variance, can be recognized as being important for establishing qualitative differences. These "small" components can be especially important for the understanding of either very small changes, or systems which possess nonlinearities.

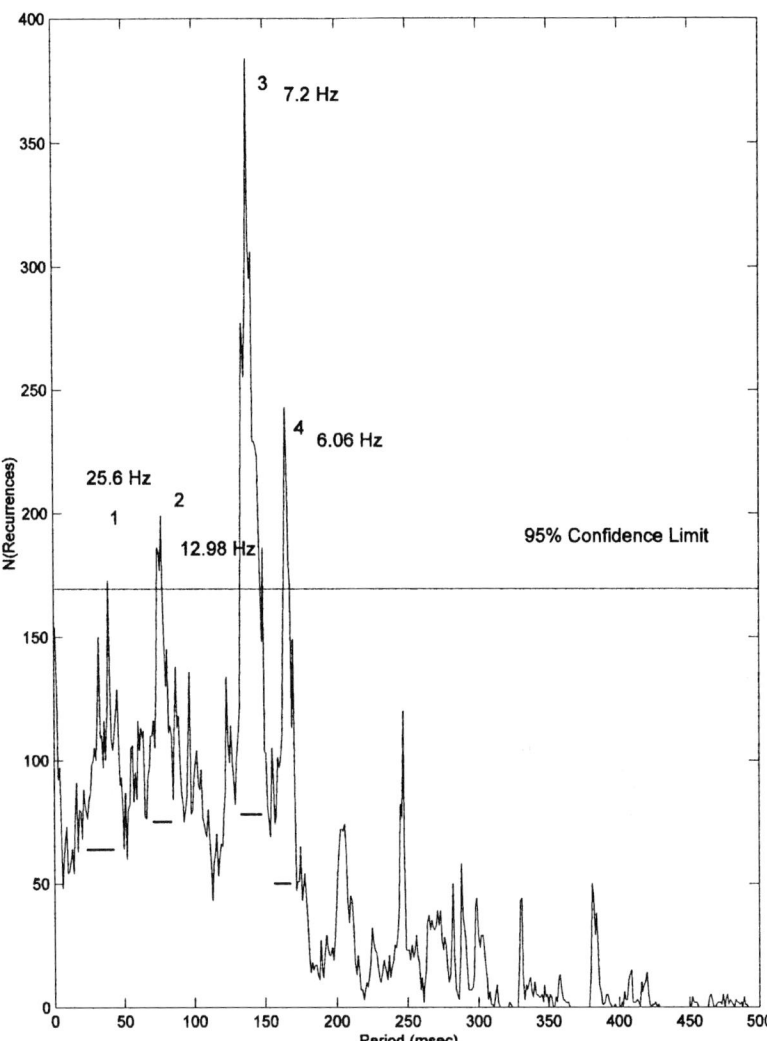

Figure 8.5. Histogram of recurrence for an episode of ventricular fibrillation of the heart. A common misconception is that the time series is chaotic. Instead, the histogram demonstrates the existence of several fairly well-defined frequencies which occur a various times through the "transient.".

8.4 Recurrence Quantification

Because graphical representation may be difficult to evaluate, RQA was developed to provide quantification of important aspects revealed by the plot. Recurrent points which form diagonal line segments are considered to be deterministic (as distinguished from random points which form no patterns). Unfortunately, beyond general impressions of drift and determinism, the plots of themselves provide no quantification. As a result, Zbilut and Webber developed several strategies to quantify features of such plots originally pointed out by Eckmann, et al. (Table 8.1)

Hence, the quantification of recurrences leads to the generation of five variables including: REC [percent of plot filled with recurrent points; i.e., $100(2q_r/(q^2-q))$, where q_r is the number of recurrent points], DET [percent of recurrent points forming diagonal lines, with a minimum of two adjacent points; i.e., $100(2q_l/(q^2-q))$, where q_l is the number of recurrent points that constitutes a line segment]; ENT [Shannon information entropy of the line length distribution; i.e., $-\Sigma\, p(l)\, \ln p(l)$ from $l=l_{min}$ to l_{max} with $p(l)$ calculated from the histogram of line lengths], MAXLINE, length of longest line segment (the reciprocal of which is an approximation of the largest positive Liapunov exponent and is a measure of system divergence); and TREND [measure of the paling of recurrent points away from the central diagonal, calculated as the regression coefficient of number of recurrent points from the diagonal line of identity perpendicularly the to plot vertex diagonal line –10% of range (to maintain reasonable statistics as the numbers used to calculate recurrences for a given indexed line decrease); i.e., $y_{rec}=c+ab$ where y_{rec} = REC for a given diagonal segment, c, a constant, a, the regression coefficient, and b, the diagonal segment index]. These five recurrence variables quantify the deterministic structure and complexity of the plot. The application of these simple statistical indexes to the recurrence plots gives rise to a five dimensional representation of the studied series. This five dimensional representation gives a summary of the autocorrelation structure of the series and was demonstrated, by means of a psychometric approach to correlate with the visual impression a set of unbiased observers derive from the inspection of a set of recurrence plots.

When one needs to appreciate eventual changes in the autocorrelation structure at the level of single element of the series, it is not possible to rely solely on the "holistic" summaries given by the direct application of RQA to the global sequence, in these cases use is made of a "windowed" version of RQA, such that for a time series $(x_1, x_2, ..., x_n)$, where ($x_j = j\tau_x$) and τ_x = sampling time.

For an n point long series

$E_1 = (x_1, x_2, ..., x_n)$
$E_2 = (x_{1+w}, x_{2+w}, ..., x_{n+w})$
$E_3 = (x_{1+2w}, x_{2+2w}, ..., x_{n+2w})$

.

.

.

$E_p = (x_{1(p-1)w}, x_{2+(p-1)w}, ..., x_{n+(p-1)w})$, (8.11)

with w = the offset, and the number of epochs (windows), E_p, satisfies the relation, $N+(p-1) \leq n$. These epochs can then individually be calculated for recurrence variables (i.e., REC, etc.), and be plotted on a time or index line. Note that the offset here is not the same as the lag. The offset refers to the original windowed data (Fig. 8.6).

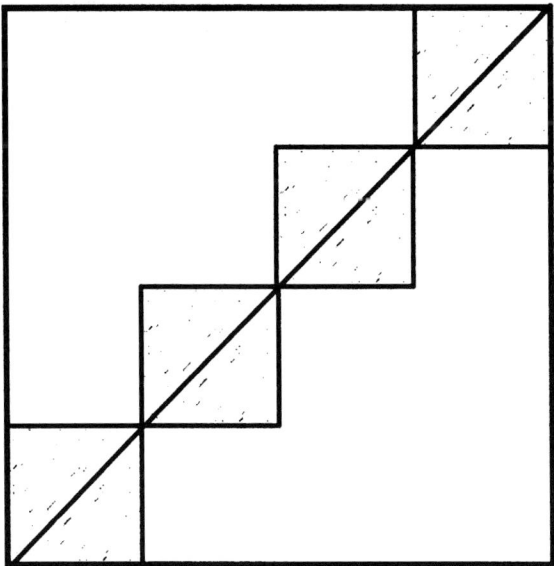

Figure 8.6. Basis for sliding window RQA: RQA variables are calculated based on small recurrences plots which may also be overlapped.

The formal mathematical definition of the RQA Measures are:

1) Rec, R, percentage of recurrence points

$$R = \frac{1}{N^2} \sum_{i,j=1}^{N} R_{i,j} . \qquad (8.12)$$

2) Determinism, *DET*, percentage of recurrence points which form diagonal lines:

$$DET = \frac{\sum_{l=l_{min}}^{N} lP(l)}{\sum_{i,j}^{N} R_{i,j}} . \qquad (8.13)$$

(P(l) is the histogram of the lengths l of the diagonal lines.)

3) Laminarity, *LAM*, Percentage of recurrence points which form vertical lines:

$$LAM = \frac{\sum_{v=v_{min}}^{N} vP(v)}{\sum_{v=1}^{N} vP(v)} \qquad (8.14)$$

(P(v) is the histogram of the lengths v of the vertical lines.)

4) Ratio, *RATIO*, ratio between *DET* and *R*:

$$RATIO = N^2 \frac{\sum_{l=l_{min}}^{N} lP(l)}{\left(\sum_{l=1}^{N} lP(l)\right)^2} . \qquad (8.15)$$

5) Averaged diagonal line length, *L*, average length of diagonal lines:

$$L = \frac{\sum_{l=l_{min}}^{N} lP(l)}{\sum_{l=l_{min}}^{N} P(l)}.$$ (8.16)

6) Trapping time, *TT*, average length of vertical lines:

$$TT = \frac{\sum_{v=v_{min}}^{N} vP(v)}{\sum_{v=v_{min}}^{N} P(v)}$$ (8.17)

7) Longest diagonal line, L_{max}, Length of the longest diagonal line:

$$L_{max} = \max(\{l_i; i = 1 \ldots N_l\}).$$

8) Longest vertical line, V_{max}, length of longest vertical line:

$$V_{max} = \max(\{v_1; l = 1 \ldots L\}).$$ (8.18)

9) Divergence, DIV, inverse of L_{max}:

$$DIV = \frac{1}{L_{max}}.$$ (8.19)

Related to the largest positive Liapunov exponent, but does not correspond to it.

10) Entropy, *ENT*, Shannon entropy of the distribution of the diagonal line lengths *p(l)*:

$$ENT = -\sum_{l=l_{min}}^{N} p(l) \ln p(l).$$ (8.20)

11) Trend, *TREND*, paling of the recurrence plot towards its edges:

$$TREND = \frac{\sum_{i=1}^{N-2}[i-(N-2)](R_i - \langle R_i \rangle)}{\sum_{i=1}^{N-2}[i-(N-2)/2]^2} \quad (8.20)$$

Table 8.1. Recurrence plot characteristics.

Feature	Interpretation
Homogeneity	stationary process
Fading to the upper left and lower right corners	nonstationarity; the process contains a trend or drift
Disruptions (white bands)	nonstationarity; some states are rare or far from the normal; transitions may have occurred
Periodic patterns	cycles; the time distance between periodic patterns (e.g. lines) corresponds to the period
Single isolated points	heavy fluctuation in the process; if only single isolated points occur, the process may be a random process
Diagonal lines (parallel to the line of identity)	the evolution of states is similar at different times; the process could be deterministic; if these diagonal lines occur beside single isolated points, the process can be from a chaotic process (if, in addition, these diagonal lines are periodic, the considered system contains unstable periodic orbits)
Diagonal lines (orthogonal to the line of identity)	the evolution of states is similar at different times but with inverse time
Vertical and horizontal lines/clusters	some states do not change or change slowly for some time (laminar states)

8.4.1 Determining Parameters for Nonstationary Series

As has been emphasized, RQA is useful for understanding nonstationary time series. Yet, since a given system may be changing state; i.e., the relevant degrees of freedom may change, the choice of m, L and r can become confounding. Unfortunately, most algorithms for such choices are based upon computer simulations of well-known, stationary examples. Typically, however, human systems are rarely stationary, and often exhibit rather sudden changes of state. Nonetheless, some guidelines can be established, based upon available research, and a careful consideration of the import of nonstationarity.

8.4.2 Choice of Embedding

In the context of nonstationarity, the notion of a "correct" embedding or delay is inappropriate. Instead it becomes important to remember that a sufficiently large embedding be chosen which will "contain" the relevant dynamics (as it may change from one dimensionality to another) as well as account for the effects of noise, which tend to inflate dimension. There are no clear guidelines relative to this question, except from what can be inferred from studies of noise. In this respect, investigators have indicated that noise will tend to require higher dimensions, even in the case of stationary dynamics. Others have studied this question in the context of a noisy Lorenz attractor, and concluded that an embedding of 6 is required to provide reasonable clarification of the dynamics. Because of the high complexity of human systems, we have empirically embedded in 10. Care, however, must be made to make sure that the system is not excessively noisy, since embedding will amplify such noise to the detriment of the real dynamics.

8.4.3 Choice of Lag

Choice of lag is governed by similar considerations. As a system changes from one dimension to another the effects of the lag are perforce changed. Thus, a so-called "optimal" lag in one embedding, becomes less so as the relevant dimension changes. Although there have been numerous proposals for choice of lag, chief among them the first local minimum of the autocorrelation or mutual information, they all are presented with the assumption of stationarity. What would appear to be more important is an understanding of how the data is acquired, as well as the system studied. For discrete data, a lag of one is usually sufficient.

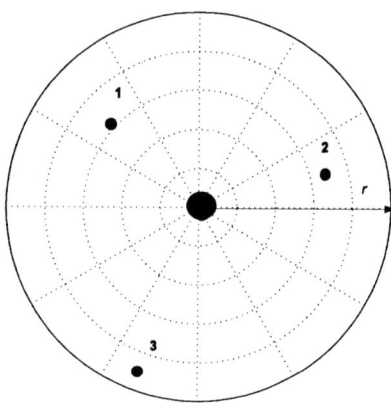

Figure 8.7. Choice of radius determines if points (1-3) are recurrent.

8.4.4 Choice of Radius

The object of RQA is to view the recurrences in a locally defined (linear) region of phase space. Practically speaking, however, because of intrinsic and extrinsic noise, too small a value of r results in quantification of the noise only; whereas too large a value captures values which can no longer be considered recurrent (Fig. 8.7). To get to the dynamics proper, a strategy is to calculate REC for several increasing values of r and to plot the results on a log-log plot to determine a "scaling" region; i.e., where the dynamics are found. Figure 8.8 demonstrates such a strategy.

If the data are extremely nonstationary, a scaling region may not be found. The guide then is the percent recurrence. A critical factor is that there be sufficient numbers of recurrences so as to make sense for computation of the other variables. A value of 1% recurrence tends to fulfill this criterion. Further verification can be obtained from an inspection of the recurrence plot: too sparse a matrix suggests a modest increase. Windowed RQA is especially informative.

If a given window fails to achieve 1% recurrence, the radius should be increased, or the window enlarged.

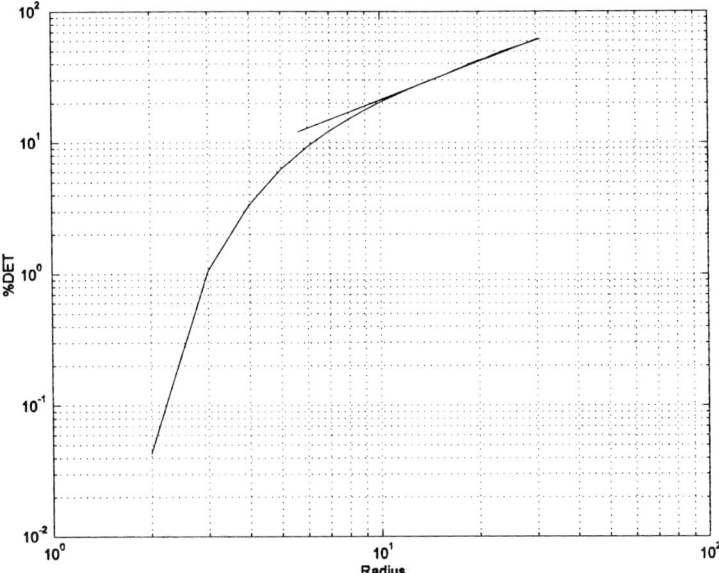

Figure 8.8. Demonstration of finding an appropriate radius. DET is calculated for several values of radius and plotted on a log-log plot. Where the line becomes scaling (here at approximately 10) is a good choice.

8.5 Detecting Singularities

Although I have argued that nondeterministic dynamics may be a better paradigm for many naturally occurring systems, an important consideration is the actual demonstration of such singularities in real data. However, part of the problem lies in modern methods of data acquisition:

Almost all experimental data currently is obtained through digitization to accommodate computer storage and manipulation. As is well known, the theory governing the sampling rate is based on linear systems theory; i.e., in order to recover all Fourier components of a periodic waveform, it is necessary to sample more than twice as fast as the highest waveform frequency v, i.e., $f_{Nyquist}=2\backslash v$. This cutoff frequency above which a signal must be sampled in order to be able to fully reconstruct it is called the Nyquist frequency. The motivation certainly, is not only that the signals of interest are captured, but also that it be done efficiently; i.e., to prevent oversampling with its increased storage requirements.

It should be noted, however, that signals which are discontinuous, nonlinear or nonstationary are not addressed by the Nuquist theorem. Additionally, real-life signals are usually prefiltered before A/D conversion stage to avoid a form of distortion known as aliasing noise as well as to limit unwanted signal components. Clearly, if a singularity were sampled it is conceivable that it would be missed given the overall sampling requirements, i.e., the sampling could be insufficient to locate it; or, prefiltering could easily distort it. To investigate these possibilities, computer and real systems have been examined with recurrence quantification, which has been demonstrated to approximate Liapunov exponents of short time series. Recurrence analysis quantifies line segment lengths derived from recurrences of a time series, which are known to be related to the largest positive Liapunov exponent.

8.5.1 *Maxline (Liapunov exponent)*

A sin wave time series was generated to approximate 10,000 Hz sampling. The signal was filtered using a bidirectional, low-pass, sixth order Butterworth filter at 2 Hz (Fig. 8.9).

Real Signal

We used the signal from the previously mentioned example of arm motion. Briefly, to recapitulate, a human arm was placed in a device, termed a manipulandum, such that the elbow was fixed and the forearm was supported by the device and allowed to move freely in the transverse plane. The subject was instructed to move the arm slowly back and forth at a comfortable pace. No pacing was suggested. The movement was digitized at 10,000 Hz in arbitrary units. The resultant activity produced a sine wave like time series (Fig. 8.10). The signals were subjected to RQA. One of the quantifications, maxline, has been shown to be approximately related to the largest positive Liapunov exponent. In a windowed form (similar to FFTs) the quantifications provide a timed view of an analyzed series.

Results

According to previous analyses, singularities should exhibit a divergence of the Liapunov exponent. Even when compared to deterministic chaotic systems where an initial volume is stretched and folded, spreading across an attractor in smooth fashion; the dynamics characterized by a singularity, scatter points randomly throughout a region of phase space once the singularity is encountered. The challenge remains for detecting putative singularities given the previously mentioned difficulties related the sampling rates. Given the automated nature of most laboratories, inattention to this concern can fail to "capture" a singularity. Additionally, an excessive concern for obtaining a "clean" signal may significantly attenuate, alter, or even obliterate a singularity by the uninformed use of digital filters. As discussed below, some degree of filtering may be allowed, but again this appears to be related to a relatively fast sampling rate.

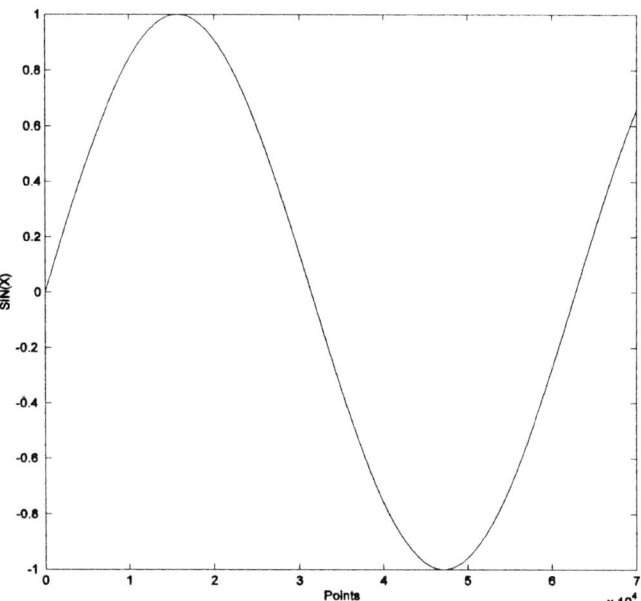

Figure 8.9. Generated sin wave.

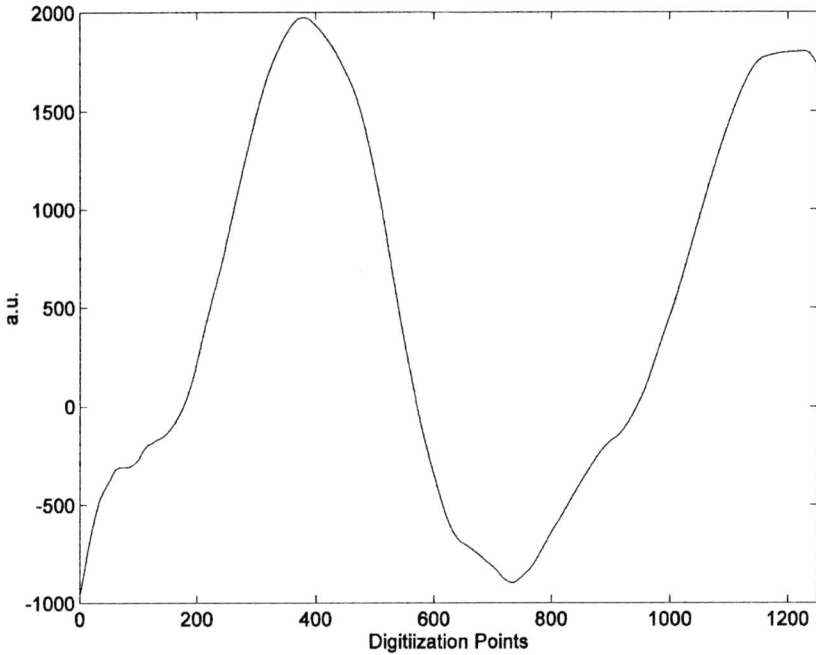

Figure 8.10. Time series generated from arm motion prior to filtering.

In contrast to the simple sin wave (not shown), the arm motion clearly demonstrates a divergence of the maxline (Fig. 8.11). This is to be expected given that at the end-point of each motion, a slight "pause" occurs, although the time series does not clearly indicate this. In fact, depending upon the sampling rate, a naïve inspection might result in the conclusion that the motion is completely smooth. The irregularities seen might be interpreted as part of the "imperfect" human motion.

The filtered form of the arm motion, however, severely attenuates the divergence (Fig. 8.12). In some sense it is remarkable that the evidence for the divergence still persists. In fact, experimentation with other types of filters and their parameters would extinguish evidence for the divergence.

In the contemporary setting use of filters is quite common—especially to avoid aliasing signals relative to sample rate. This evidence, however, points to the need to reconsider such a strategy. If nondeterministic, or non-smooth dynamics are entertained, signal sampling should be governed by the highest frequency possible without filtering. (It is recognized that band limiting is, in fact, a form of filtering—the idea is to make sure no small nonlinearities are missed.) If such discontinuities are found, then experimentation with downsampling can be carefully entertained. Prior "automatic" or adaptive methods in the face of new phenomena may suggest an artificial view of the dynamics.

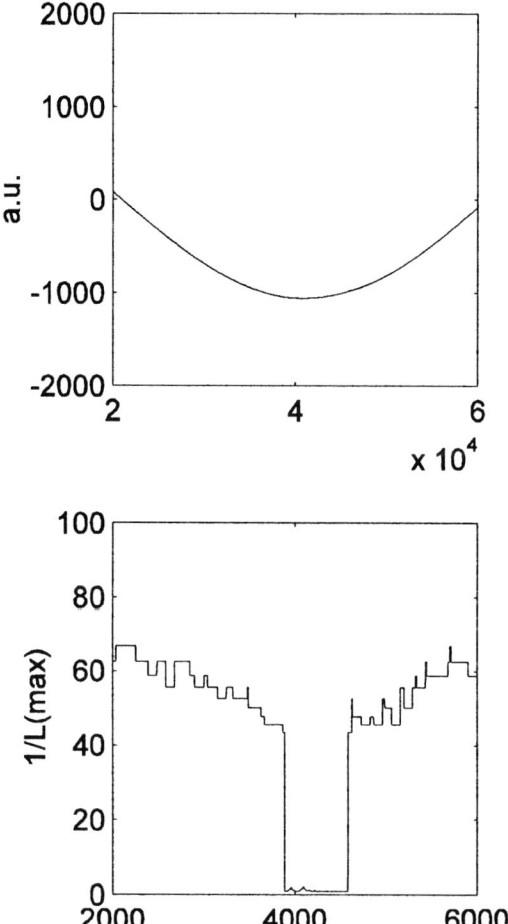

Figure 8.11. Detail of arm motion (top) and divergence of maxline (1/L(max) on the bottom.

202   *Unstable Singularities*

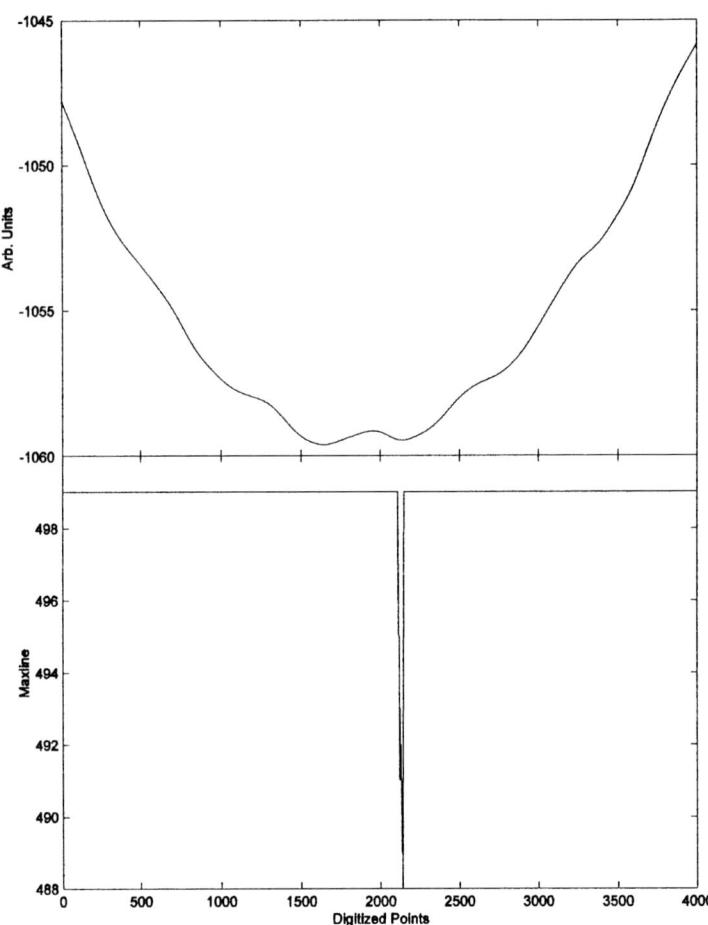

Figure 8.12. Recorded divergence after signal filtering. The maxline divergence is still present, however, it has been severely attenuated. In some cases, increasing the order of the filter, or changing the filter type, resulted in a total elimination of the divergence. It did, though, result in a smoother, "cleaner" signal. In this case a digital bidirectional Butterworth filter was employed

An example of the difficulty can be appreciated from the time series of the neutron star equations. Figure (8.13) demonstrates deflections not visible, but clearly indicated by the maxline variable

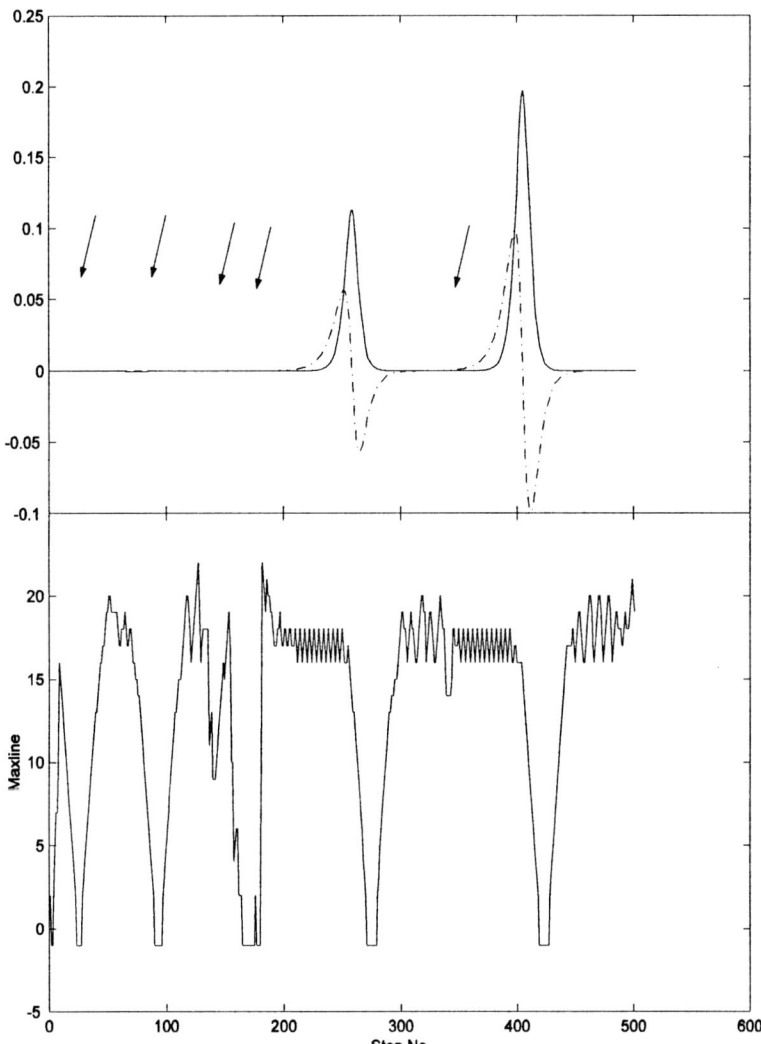

Figure 8.13. Neutron star equations (top) with resultant calculation of maxline (bottom). Arrows indicate the presence of very small deflections not easily visible, but are clearly indicated by the maxline variable. Many more smaller deflections are present, however, this would swamp the graphic with arrows. These are indicated by the very small oscillations of maxline.

Batch Reactions

The utility of this approach can be more than theoretical interest, as the following example from chemical reactions demonstrates.

In spite of constant attempts to maintain safety in chemical processes, there is always the possibility of "runaway" reactions which can be dangerous. Consequently, considerable resources have been expended in maintaining "early warning" detection systems.

A common method for monitoring reactions is to use the second derivative of some system variable such as temperature. As is well appreciated by most experimentalists, however, use of this criterion can be problematic, especially in the case of noisy data.

In a series of works Strozzi, and Zaldívar and their colleagues have suggested a new criterion to delimit runaway boundaries, using nonlinear dynamical theories and the fact that the sensitivity to initial and operating conditions is a well-known characteristic of chemical reactors. According to the analysis and previous criteria, the early warning detection criterion was defined as when the divergence of the reactors becomes positive on a segment of the reaction path, i.e. $div > 0$. It is recalled that the divergence is a scalar quantity defined at each point as the sum of the partial derivatives of the mass and energy balances with relation to the corresponding variables, temperature and conversions; i.e., $\partial (dT/dt)/\partial T + \sum \partial(dz_i/dt)/\partial z_i$. This criterion was compared with previous ones. The results show its validity even in the case of autocatalytic reactions, where previous criteria were unable to define a suitable boundary for runaway characterization. It has been extended to other type of reactions, i.e. parallel, consecutive; other type of reactors, i.e. batch, semibatch and continuous stirred tank reactors, and different operating conditions, i.e. isoperibolic—constant jacket temperature—and isothermal—with controlled jacket temperature.

Furthermore, Stozzi showed, using simulated data, that the divergence of the system could be reconstructed from only temperature measurements and that by measuring several temperatures inside the reactor one could do the same experimentally. However, the procedure was not sufficiently robust against noise in temperature measurements for the pilot plant (100 L) reactor experiments. An improvement using a related technique, i.e. RQA, was developed in which only one temperature measurement is needed and which is robust against noise contamination. However, despite the fact that there is a relation between the values obtained in the RQA and the divergence of the system, it is difficult to quantify this relation.

An isoperibolic batch reactor system can be modeled as:

$$\frac{du_A}{dt} = -f_1 u_A^{n_1}, \qquad (8.21)$$

$$\frac{du_B}{dt} = f_1 u_A^{n_1} - \rho f_2 u_B^{n_2}, \qquad (8.22)$$

$$\frac{d\theta}{dt} = \alpha f_1 u_A^{n_1} + \alpha \lambda \rho f_2 u_B^{n_2} - \beta(\theta - 1). \qquad (8.23)$$

The reactions are assumed to be n_i-th order with respect to reactant and to follow Arrhenius temperature dependence, i.e. $f_i = \exp[\gamma_i (\theta - 1)/\theta]$. ρ and λ are the reaction rate constant ratio and the heat of reaction ratio ($\Delta H_2/\Delta H_1$), respectively; whereas α and β are the dimensionless heat of reaction and dimensionless heat transfer parameters (see Notation for a complete definition of all variables and parameters).

The phase space reconstruction theory has been developed for studying the attractor that will be reached by our system, i.e. the long time behavior of the system. However, in the case of batch and semibatch reactors, the final attractor is a fixed point, i.e. when all reactants have been depleted and the reactor temperature is in equilibrium with the jacket temperature, which has not so much interest. In these reactors, the interesting region in which we are tying to reconstruct the system is the transient region

Let us consider a system of n ordinary differential equations, in our case energy and mass balances, defined as:

$$\frac{d\mathbf{x}(t)}{dt} = \mathbf{F}(\mathbf{x}(t)) \qquad (8.24),$$

where $\mathbf{x}(t) = [x_1(t), x_2(t),..., x_n(t)]$ in R^n and $\mathbf{F} = [F_1,..., F_n]$ is a smooth nonlinear function of $\mathbf{x}$, i.e. that the existence and uniqueness properties hold. At time $t > 0$ the initial condition $x(0)$ finds itself at some new point $x(t)$. Similarly, all initial conditions lying in a certain region $\Gamma(0)$ find themselves in another region $\Gamma(t)$

after time t. If we let $V(t)$ denote the volume of the region $\Gamma(t)$, then a strong version of Liouville's theorem states that:

$$\frac{dV(t)}{dt} = \int_{\Gamma(t)} tr[J(x)]dx_1...dx_n \quad (8.25),$$

where

$$div[J(x)] = \frac{\partial F_1(x)}{\partial x_1} + ... + \frac{\partial F_n(x)}{\partial x_n}, (8.26)$$

and J is the Jacobian (a function used in the calculation of derivatives and divergence of vector fields.). Assuming that our n-dimensional volume is small enough that the divergence of the vector field is constant over $V(t)$,

then

$$\frac{dV(t)}{dt} = V(t) \cdot div[J(x)], \quad (8.27)$$

and hence,

$$\int_0^t \frac{dV(\tau)}{V(\tau)} = \int_0^t div[J(x)]d\tau, \quad (8.28)$$

which means that the initial phase space volume $V(0)$ shrinks (grows) with time in R^n as:

$$V(t) = V(0) \cdot \exp\left[\int_0^t div[J(x)]d\tau\right]. (8.29)$$

Mathematical Appendix

Hence, for the case of a system given by the equation, the rate of change of an infinitesimal volume $V(t)$ following an orbit $x(t)$ is given by the divergence of the flow which is locally equivalent to the trace of the Jacobian. The integral of a strictly positive (negative) function is itself strictly positive (negative), and the integral of an identically zero function is identically zero. That means if $\text{div}[\mathbf{F}(\mathbf{x})] < 0$ in the state space then the flow of trajectories is volume-contracting, if $\text{div}[\mathbf{F}(\mathbf{x})] > 0$ the flow is volume-expanding, and if $\text{div}[\mathbf{F}(\mathbf{x})] = 0$ then the flow is volume-preserving.

There are several methods to calculate numerically the divergence of a system. In principle the results should be equivalent, but due to numerical truncation errors there are several differences in the numerical results. However, all these methods are based on the general principle that there are a set of nearby trajectories and that between this set the closest ones are selected to evaluate how the system evolves. These are applied in the transient phase, so several close trajectories are necessary, which, in principle, may be generated numerically by changing slightly the initial conditions from the beginning of the integration procedure or at each time step.

However, if it is assumed that the time step from one point to another in the time series is short enough that the Jacobian of the system has not substantially changed, then it should be possible to use only one trajectory, by assuming that to pass from $\delta x(i)$ to $\delta x(i+h)$, $\mathbf{J}$ is used to pass from $\delta x(i)$ to $\delta x(i+2h)$, $\mathbf{J}^2$, to pass from $\delta x(i)$ to $\delta x(i+3h)$, etc.

One way to calculate the divergence from numerical/experimental time series is to reconstruct directly the Jacobian and to calculate its trace. The time evolution of infinitesimal differences is completely described by the time evolution in tangent space, i.e. by the linearized dynamics. Let $\mathbf{x}^{(1)}$ and $\mathbf{x}^{(2)}$ be two such nearby trajectories in m-dimensional state space. Considering the dynamical system as a map, the time evolution of their distance is

$$\mathbf{x}_{n+1}^{(1)} - \mathbf{x}_{n+1}^{(2)} = \mathbf{F}(\mathbf{x}_n^{(1)}) - \mathbf{F}(\mathbf{x}_n^{(2)}) = \mathbf{J}_n(\mathbf{x}_n^{(1)} - \mathbf{x}_n^{(2)}) + O(\|\mathbf{x}_n^{(1)} - \mathbf{x}_n^{(2)}\|^2)$$

(8.30)

where we have expanded $\mathbf{F}(\mathbf{x}_n^{(1)})$ around $\mathbf{x}_n^{(2)}$ and $\mathbf{J}_n = \mathbf{J}(\mathbf{x}_n^{(2)})$ is the $m \times m$ Jacobian matrix of $\mathbf{F}$ at $\mathbf{x}^{(1)}$.

For the case of a discrete system –experimental data points- it is possible to transform the linear equation, which represents the Jacobian evolution: $\dot{\mathbf{y}} = \mathbf{J} \cdot \mathbf{y}$ into $\mathbf{y}(t + \Delta t) = (\Delta t \cdot \mathbf{J} + \mathbf{I}) \cdot \mathbf{y}(t)$ and then to calculate $\mathbf{J}$ as a function of neighbor trajectories at each time step, whereas $\mathbf{y} = \delta\mathbf{x}$.

In order to calculate the Jacobian matrix, we need to solve a linear system equations given by the best linear fit of the map which for $\mathbf{x}_n^{(1)}$ close to $\mathbf{x}_n^{(2)}$ sends $\mathbf{x}_n^{(1)} - \mathbf{x}_n^{(2)}$ to $\mathbf{x}_{n+1}^{(1)} - \mathbf{x}_{n+1}^{(2)}$. Close here means that the map should be approximately linear. Depending on the number of close trajectories, p, and the dimension of the system, m, we may have several situations. In the case of $p < m + 1$ the system is undetermined, if $p = m + 1$ it is possible to find an exact solution, and when $p > m + 1$ the system is overdetermined and we may find a least squares solution.

An important consideration here is the fact that it is assumed that the system is smooth and that unique solutions exist. However, Stozzi has pointed out that noisy data makes this system difficult to work with. Returning to the arguments that Dave Dixon suggested previously in the neutron star equations, the noise destroys the possibility of smooth dynamics. Indeed, in real reactions temperature gradients can develop, and it can be argued that at some points small temperature anomalies can develop into a singular blow up which results in a runaway.

Experiments

So-called "runaway" and "non-runaway" esterification experiments were performed to test the idea that recurrence plot-based PLEs could effecitvely monitor the reactions as a "proof" of concept. The experiments were based on typical reactions performed on large-scale bases in industrial settings, and are a constant threat to safe operation, redounding also to concerns for environmental pollution. The term "runaway" refers to loss of control of the process to the point that the temperature increases suddenly with possible explosive results Traditionally, the divergence in such a scenario has been monitored by the second derivative. As many experimentalists are aware, however, the second derivative very easily amplifies any noise, such that it becomes almost impossible to distinguish significant changes in the dynamics. From a practical point of view, one can easily see that such a scenario poses a significant dilemma for an observer monitoring such reactions—should a perceived change be regarded as dangerous and the reaction terminated, with considerable waste of time and resources—or is the change simple illusory?

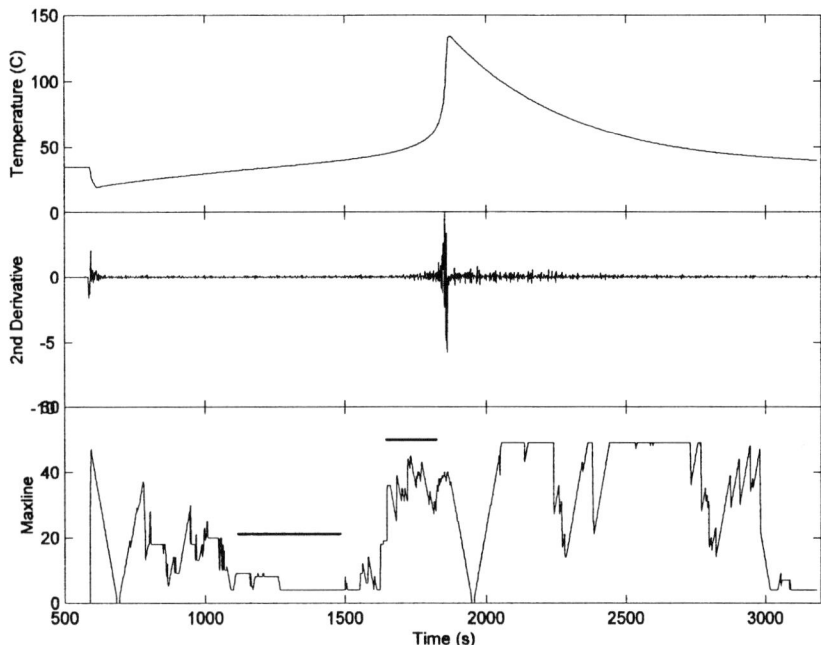

Figure 8.14. Runaway reaction with constant jacket temperature. Note the change in maxline (overstrike) at 1000 s indicating change, followed by stabilization (2nd overstrike), and then a precipitous drop.

Considerable effort has been expended to devise methods to provide early warning for such events. The objective, of course, is not only to improve safety, but also to reduce costly false positive alarms. The experiments were performed in "batch" (reactants added simultaneously), and "semibatch" (reactant added during the experiment at a constant rate) conditions. One runaway experiment was performed under "isothermal" conditions; i.e., an attempt was made to control the temperature inside the reactor by adjustment of the reactor jacket temperature. Additionally, intermediate profiles were carried out, as well as controlled experiments. In all cases the variable of interest was temperature as monitored by sensors within the reactors. A 50 or 100 point window was chosen to obtain optimal sensitivity. Thus, the maximal line segment measurable depends upon this window length criterion.

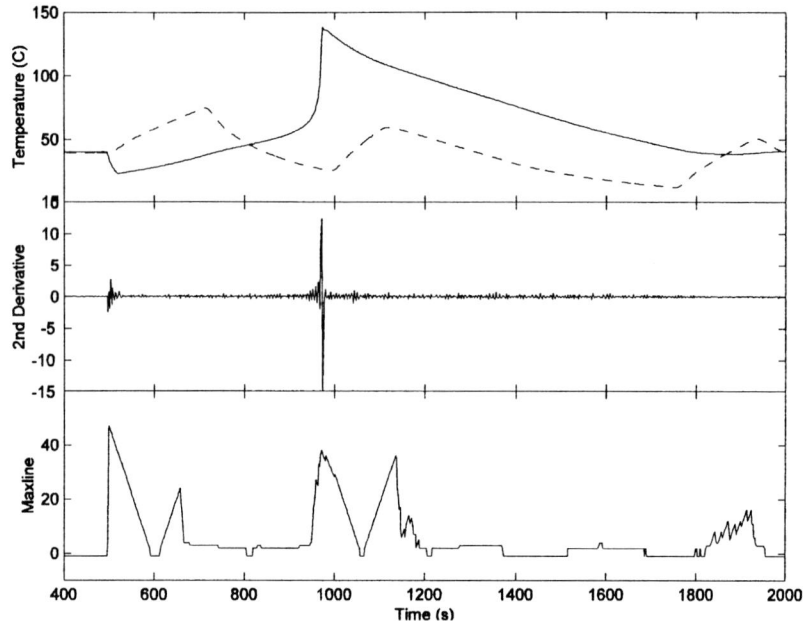

Figure 8.15. Batch runaway reaction with attempt at control by changing jacket temperature. Dashed line denotes jacket temperature.

Results

What was noteworthy was that most of the data had minimal noise, such that the 2nd derivative could easily register a transition. (The maximal noise remained constant across the different experiments as registered by the 2nd derivative.) Important also is the interpretation of the maxline variable relative to the dynamics: recall that a positive Liapunov exponent suggests instability. In this case, a large maxline indicates less instability than a smaller one (albeit it is still instability), relative to the local (windowed) dynamics. Thus, the maxline cannot inform an observer if the dynamics are "dangerous—only that they are changing. And it is the direction of the measured time series that defines such a condition.

Figure 8.14 depicts the results of a batch runaway experiment under isoperibolic conditions. Clearly the temperature increases, however, the second derivative becomes significant (registering a large divergence) only at the last moment. The maxline describes an interesting scenario: there appears to be some noise-based variation until approximately 1100 s, at which point, the value becomes very low indicating (relatively) high instability proceeding until approximately 1600 s (first solid bar). At this point, the maxline increases and stays approximately the same for the next few hundred seconds (next solid bar). The interpretation may follow along these lines: shortly after the beginning of the reaction, the temperature begins to rise slowly (in an unstable way); i.e., no clear indication of runaway. However, it suddenly "stabilizes" in this temperature increase because the deterministic recurrent points "pile-on" quickly because of its acceleration. After 2000 s the temperature decelerates in a stable way (punctuated by some noise). In contradistinction to this, the 2nd derivative registers a change only at the last few seconds as the temperature maximum is achieved. In Fig. 8.15, a batch runaway (2 L reactor) cannot be controlled by changing jacket temperature.

The second derivative momentarily registers a small change at the beginning, and again just before the maximum. The maxline, however, is sensitive initially to the increasing temperature (low maxline), as well as, apparently, to the subtle changes involved in changing jacket temperature. This is contrasted with Fig. 8.16 which depicts a simple experiment without a reaction. Instead, the jacket temperature controls the reactor temperature. Here there is no subtle increase in temperature, and in the acceleration phase, the maxline is very uneven and promptly registers the fact that the reactor temperature has stabilized. Note that there is no maxline even indicated (the value is -1, a flag indicating no line segments). Figure 8.17 is similar to the process in Fig. 8.14 except that this is a semibatch experiment, and the reactor was much larger (100 L). Thus the method is robust with rescaling.

Figures 8.18, 8.19 present the results for intermediate profiles of semibatch experiments in 100 L reactors. What is clearly of interest is that in neither of these experiments, does the second derivative hint of any accelerating process. No doubt this is due to the control of the process at the last few seconds. However, if the maxline criterion is invoked, the interpretation could easily be that runaway was incipient. Clearly, this experiment was contrived to create a scenario of runaway controlled at the last minute. Note also that as the processes change (decrease temperature), small maxline increases are registered

Finally, Fig. 8.20 demonstrates results in a controlled reaction (semibatch). The maxline briefly increases, is not sustained, and quickly dies out, indicating absence of runaway.

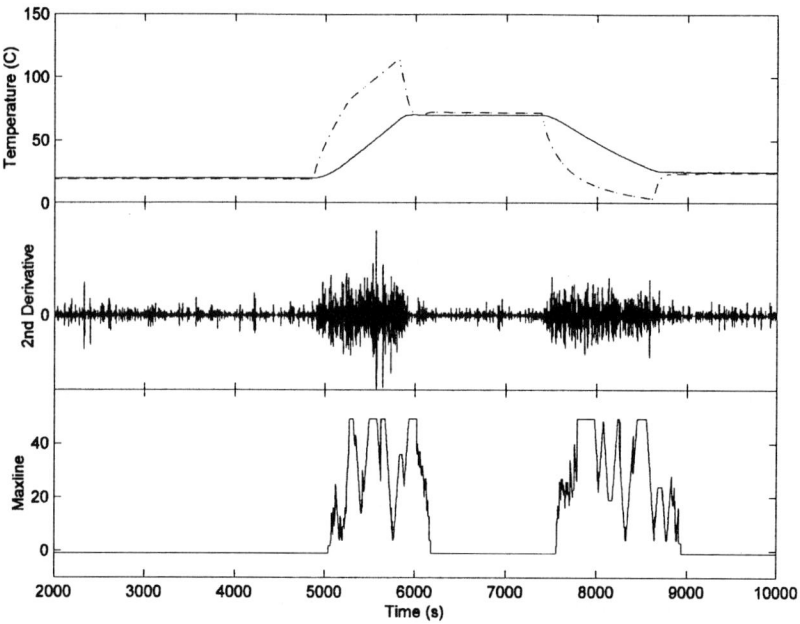

Figure 8.16. Semibatch runaway experiment. No reaction. Operator manipulation in heating/cooling experiment. (Dashed line denotes jacket temperature.)

As can be seen from these examples, the existence of singularities can be inferred from a good working knowledge of the dynamical process, and substantiation from RQA maxline calculations. The former requisite needs to be emphasized, since maxline is not specifically indicative of singularities.

8.5.2 Orthogonal Vectors

Another method to detect singularities is to search for orthogonal vectors. By orthogonal we mean forming an angle of 90 degrees with respect to each other, which is a sign of independence. This approach can be useful for large amplitude oscillations as are found in the example below.

For a simple oscillation (Figs. 8.21, 8.22), the vector time series is plotted against a delayed version of itself. At the apex (if there is a singularity) the vectors are found to be orthogonal to each other; i.e., they form a right angle. In effect, this is equivalent to identifying a shift in the time line of a recurrence plot.

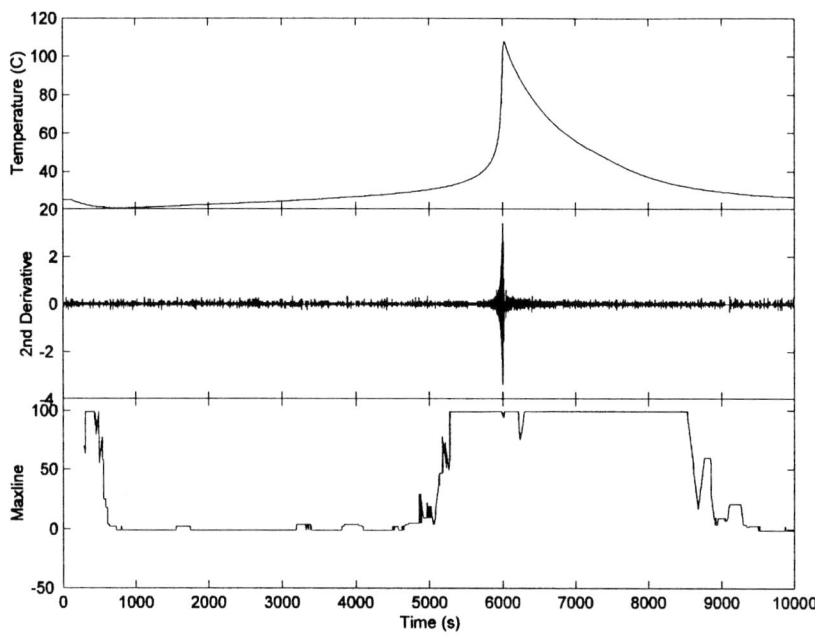

Figure 8.17. Semibatch runaway experiment in 100 L reactor.

8.5.3 Some Observations

Given the difficulty of determining singularities, it is natural to question the utility of knowing of their existence. Consideration of the arm motion is instructive. Although the basic deterministic pattern is known (i.e., the sine-like waveform), technically, the pattern is unpredictable due to the variable "pause" of the singularity. Indeed the long-term future motion can only be described in a statistical sense. This is due to the "explosion" of information at the singular point: because of the variability of the pause, a specific trajectory cannot be determined by any effort of modeling. The scenario is similar for the simple harmonic oscillator (SHO) previously discussed. At the tangent, any number of circles may develop—predictability again is only possible in a statistical sense. Thus determination of the existence of a singular point is crucial to the understanding of the dynamics. Failure to do so gives a completely misinformed understanding of the dynamics.

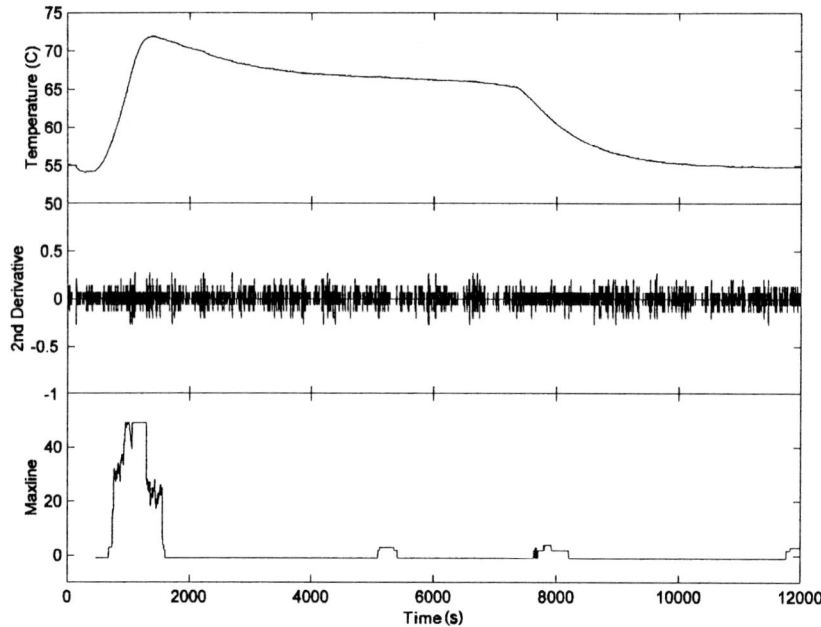

Figure 8.18. Intermediate profile of semibatch experiment. Initial runaway is controlled.

In these examples (including the reactions), it should be underscored that as with all experimental algorithms involved in the calculation of positive Liapunov exponents and orthogonal vectors, there is no comparison of the phase-space volume. This is especially obviously the case with nonstationary, transient data. Thus, if there is a change in the exponent, no evaluation can be made to determine multidimensional volume direction. To be useful, then, it is important to track simultaneously the direction of the variable of interest.

Related to this last point is the question of significance of the maxline changes. Clearly some noise can trigger maxline changes, and can present a false impression of the dynamics. Since these examples were designed as "proof of concept," no clear criteria were developed to determine level changes of maxline to suggest singularities. This was to recognize that real conditions would be scaled up and require further evaluation. Nonetheless, previous experience with other systems suggest that an appropriate algorithm is feasible. Specifically, baseline 95% confidence limits for maxline can be obtained and used as a boundary for significant increases/decreases and compared to variable's direction

change

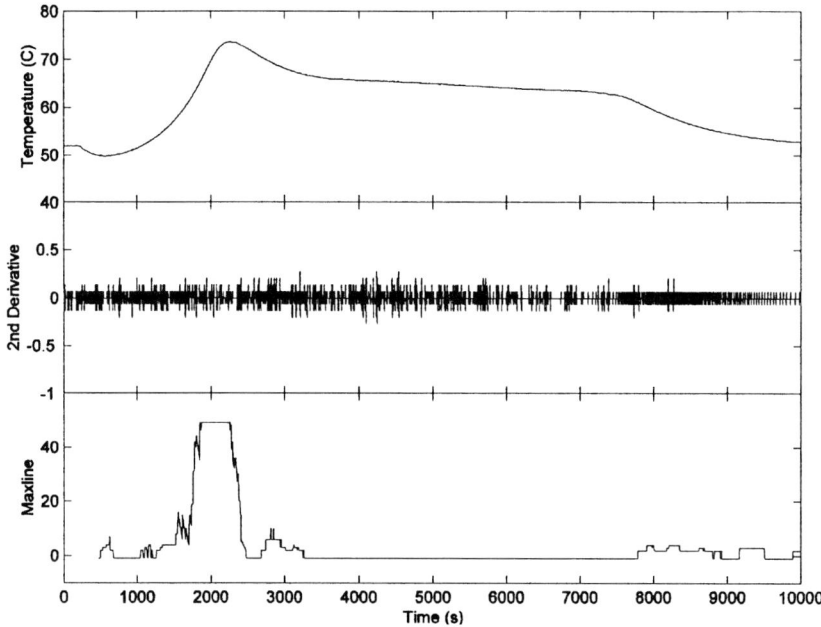

Figure 8.19. As in the previous figure, but runaway is greater.

Finally, although, numerous studies have attempted to document that a system is chaotic, it should be recalled that positive Liapunov exponents have a separate utility; namely, to determine the stability of a system. Although the term, "stability," tends to be used primarily in the physical sciences and mathematics, biologists should consider the term in conjunction with a "state" change.

The previous examples have attempted to document this utility by using a simple recurrence plot-based algorithm which can be implemented on-line. Clearly, other systems which are typically noisy and nonstationary, such as biological, financial and physical may also profit from such analyses.

Frequently, it can be difficult to determine if a specific "jiggle" in a time series is really indicative of a significant state change, versus either a minor fluctuation or noise—or, for that matter, if a noise is significant to change the dynamics. (Are the dynamics being affected by noise in the system, or is the noise external to the dynamics in the sense that it is due to the recording system.)

Although these methods may not give a definitive answer, it may provide sufficient objective evidence to make a reasonable conclusion based upon an adequate familiarity with the dynamics.

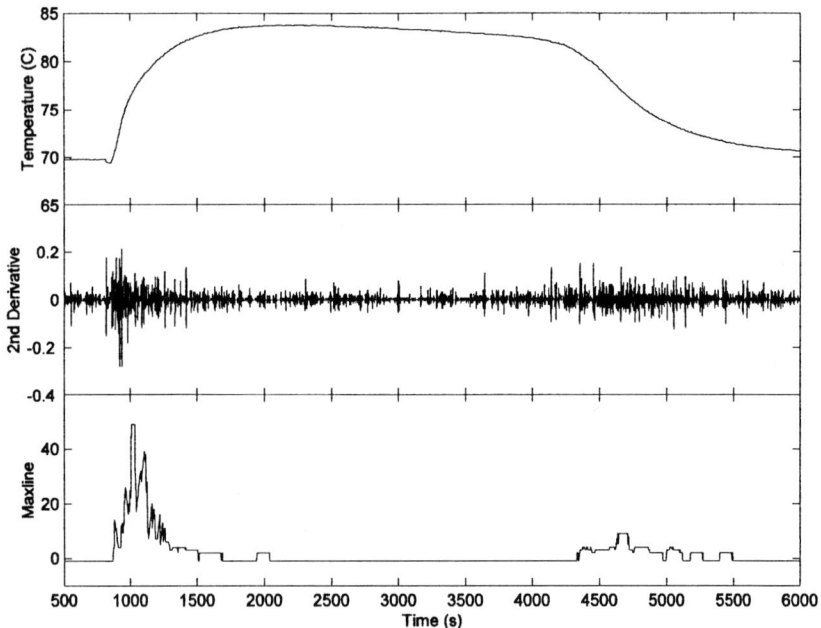

Figure 8.20. Controlled semibatch reaction.

Mathematical Appendix 217

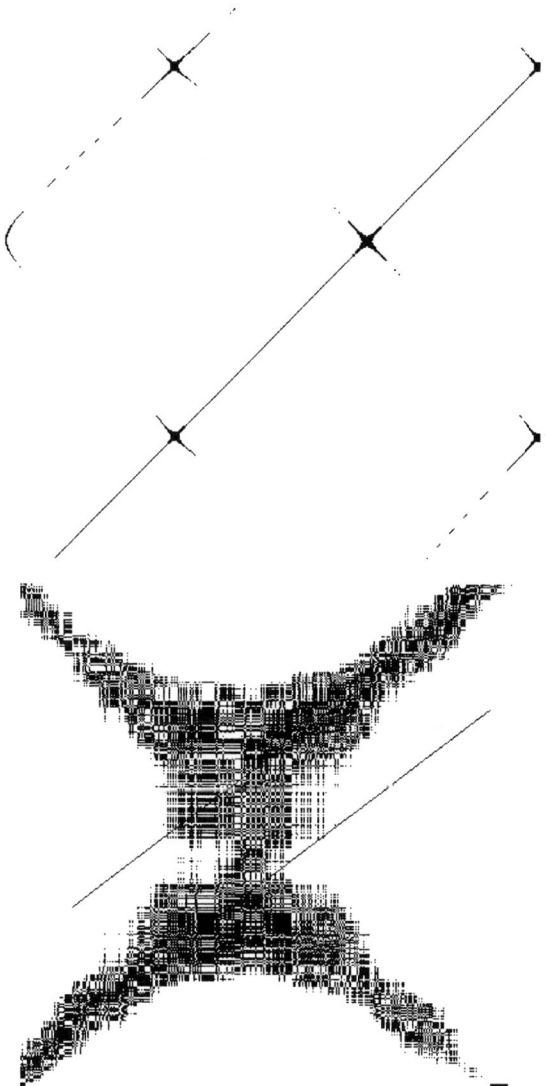

Figure 8.21. Barely visible shift of trajectories in recurrence plot (top), and detail (bottom) showing shift in trajectory—equivalent to orthogonal vectors. (Diagonal lines mark the primary trajectories before and after the singularity.)

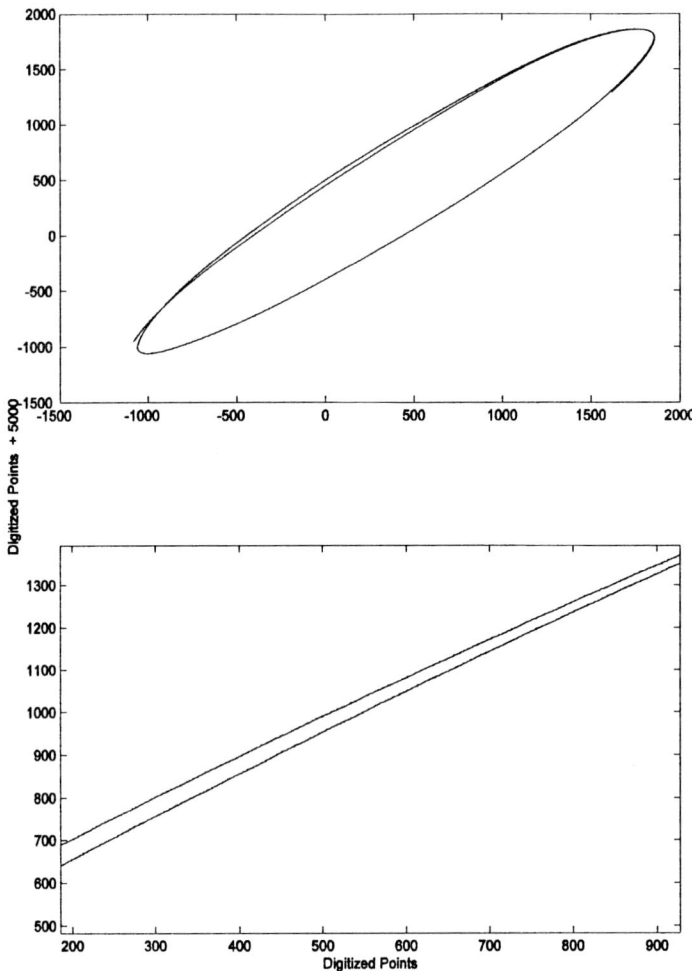

Figure 8.22. Orthogonal vectors (perpendicular to each other) are seen through a plot of the original vector (top) vs. a delayed version (bottom, right). This is seen through the "squaring off." Indeed, the view is not clear. This technique is not optimal since it requires "trial and error" to determine an appropriate time delay which would allow for adequate visualization of the independent vectors. Even if it seems that there is an optimal delay and visualization, the situation may not be sufficiently clear. An initial guess can be obtained by using the autocorrelation function or the mutual information (as previously discussed). With initial high rates of sampling, however, this still will result in many repetitive acts. In

practice, the recurrence plot with several careful refinements as seen in Figure 8.21 may prove to be the most facile method.

Bibliography

The bibliography is intended to refer to works dealing with singularities and related ideas and techniques, as well as peripheral considerations.

Books

Arfken GB, Weber HJ (1995). *Mathematical Methods for Physicists.* Academic Press, San Diego.

Arnold VI (1988). *Mathematical Methods of Classical Mechanics.* Springer, Berlin.

Bak P (1996). How Nature Works: The Science of Self-organized Criticality. Copernicus Press, New York.

Barrrett W. (1978). *The Illusion of Technique.* Anchor Press/Doubleday, Garden City, New York.

Brockwell PJ, Davis RA (1991). *Time Series: Theory and Methods.* Springer, New York.

Campbell L, Garnett W (1884). The Life of James Clerk Maxwell, with Selections from his Correspondence and Occasional Writings. MacMillan and Co, London.

Coddington EA, Levinson N (1955). *Theory of Ordinary Differential*

Equations. McGraw-Hill, New York.

Csikszentmihalyi M (1996). *Creativity.* HarperCollins, New York.

Einstein A (1983). *Ather und Relativitats—Theorie.* [Sidelights on relativity]. Dover, New York.

Feller W (1968). *An Introduction to Probability Theory and its Applications.* Wiley, New York.

Garnett CL. (1884). *The Life and Times of James Clerk Maxwell, With Selections from his Correspondence and Occasional Writings.* London, MacMillan and Co.

Hebb DO (1949). *The Organization of Behavior.* Wiley, New York.

Kitney RI, Rompelman O (1980). *The Study of Heart Rate Variability.* Clarendon Press, Oxford.

Landau LD, Lifshitz EM (1959). *Fluid Mechanics.* Vol 6. *Course of Theoretical Physics.* Pergamon Press, London.

Lovejoy AO (1971). *The Great Chain of Being.* Harvard University Press, Cambridge (originally published 1936).

Müller B, Reinhardt J (1990). *Neural Networks.* Springer, Berlin.

Popper K (1959). *The Logic of Scientific Discovery.* (trans. of *Logik der Forschung*). Hutchinson, London.

Prigogine I (1980). *From Being to Becoming.* Freeman & Co., San Francisco.

Sartre J-P, Alexander L (trans.) (1976). *Nausea.* New Directions, New York.

Schroeder M (1991). *Fractals, Chaos, Power Laws.* WH Freeman and Co., New York.

Shannon C, Weaver W (1949). *The Mathematical Theory of Communication.* University of Illinois Press, Chicago.

Whitney CA (1990). *Random Processes in Physical Systems.* Wiley & Sons, New York.

Winfree AT (1987). *When Time Breaks Down.* Princeton University

Press, Princeton.

Zak M, Zbilut JP, Meyers RE (1997). *From Instability to Intelligence: Complexity and Predictability in Nonlinear Dynamics. (Lecture Notes in Physics:* New Series m 49). Springer, Berlin Heidelberg New York.

Book Chapters, Proceedings

Crutchfieled JP (1993). Observational complexity and the complexity of observation. *Inside vs. Outside.* Atmanspacher H (ed), Springer, Berlin, pp 234-273.

Haken H (1991). Synergetics—can it help physiology? *Rhythms in physiological systems.* Haken H, Koepchen (eds), Springer, Berlin, pp 21-31.

Keegan AP, Zbilut JP, Merritt SL, Mercer PJ (1993). Use of recurrence plots in the analysis of pupil diameter dynamics in narcoleptics. *SPIE Proceedings: Chaos in Biology and Medicine.* Vol 2036, pp 206-213.

Mumford D (2000). The dawning of the age of stochasticity. *Mathematics: Frontiers and Perspectives.* Atiyah AM, Lax M, Mazur B (eds), American Mathematical Society, Providence, RI, pp 197-218.

Santucci P, Zbilut JP, Mitra R (1998). Detecting undersensing in implantable defibrillators using recurrence analysis. *Computers in Cardiology 98.* 261-264.

Webber Jr CL, Zbilut JP (1994). Neural net estimation of cardiac nondeterminisim. *Intelligent Engineering Systems Through Artificial Neural Networks.* Vol 4. ASME Press, NY, pp 695-700.

Webber CL, Jr, Zbilut JP (1996). Assessing deterministic structures in physiological systems using recurrence plot strategies. *Bioengineering Approaches to Pulmonary Physiology and Medicine.* Khoo MCK (ed), Plenum Press, N Y, Chapter 8, pp 137-148, 1996.

Webber, Jr. CL and Zbilut JP (1998). Recurrent structuring of dynamical and spatial systems. *Complexity in the Living: A Modelistic Approach*. Colosimo A (ed). Universita di Roma "La Sapienza", Rome, pp 101-133.

Zbilut JP, Mayer-Kress G (1988). Dimensional analysis of heart rate variability in heart transplant recipients. *Proceedings of Conference on Nonlinear Dynamics in Medicine and Biology. Mathematical Biosciences* 90(1-2): 49-70.

Zbilut JP, Hübler A, Webber Jr CL (1996). Physiological singularities modeled by nondeterministic equations of motion and the effect of noise. *Fluctuations and Order: The New Synthesis.* Millonas M (ed). Springer, New York, pp 397-417.

Zbilut JP, Santucci PA, Yang S-Y, Podolski, JL (2002). Linear and Nonlinear Evaluation of Ventricular Arrhythmias. *Medical Data Analysis: Proceedings of the Third International Symposium.* Colosimo A, Giuliani A, and Sirabella P (eds). (Lecture Notes in Computer Science 2526). Springer, Berlin.

Zbilut JP, Webber Jr., CL, Zak M (1998). Quantification of heart rate variability using methods derived from nonlinear dynamics. *Assessment and Analysis of Cardiovascular Function.* Drzewiecki G, and Li J K-J, (eds.). Springer, New York, Chapter 19, pp 324-334.

Zbilut JP, Zak M, Webber, Jr CL (1994). Nondeterministic chaos approach to neural intelligence. *Intelligent Engineering Systems Through Artificial Neural Network.,* Vol 4. ASME Press, NY, pp 819-824.

Journals

Bak P, Tang C, Wiesenfeld K (1988). Self-organized criticality. *Physical Review A* 38: 364-374.

Balocchi R, DiGarbo A, Michelassi C, Chillemi S, Varanini M, Barbi M, Legramante JM, Raimondi G, Zbilut JP (2000). Heart rate and blood pressure response to short-term head-down bed rest: a nonlinear approach. *Method Inform Med* 39: 157-159.

Barnes B, Grimshaw R (1997). Analytic and numerical studies of the Bonhoeffer Van der Pol system. *J Austral Math Soc Ser B* 38: 427-453.

Bradley E, Mantilla R (2002). Recurrence plots and unstable periodic orbits. *Chaos* 12: 596-600.

Chen Z-Y (1990). Noise-induced instability. *Phys Rev A* 42: 5837-5843.

Censi F, Calcagnini G, Bartolini P, Bruni C, Cerutti S (2002). Spontaneous and forced non-linear oscillations in heart period: role of the sino-atrial node. *Medical Engineering & Physics* 24: 61-69.

Chen Z.-Y. Noise-induced instability. *Physical Review A* 42: 5837-5843.

Crutchfield JP, Kaneko K (1988). Are attractors relevant to fluid turbulence? *Physical Review Letters* 60: 2715-2718.

Dobson CM (2003). Protein folding and disease. *Nat Rev Drug Discov* 2: 154-160.

Dippner JW, Heerkloss R, Zbilut JP (2002). Recurrence quantification analysis as a tool for characterization of non-linear mesocosm dynamics. *Marine Ecology Progress* 242:29-37.

Dixon DD (1995). Piecewise deterministic dynamics from the application of noise to singular equations of motion. *Journal of Physics A* 28: 5539-5551.

Dixon DD, Cummings FW, Kaus PE (1993), Continuous "chaotic" dynamics in two dimensions. *Physica D* 65: 109-116.

Eckmann JP, Kamphorst SO, Ruelle D (1987) Recurrence plots of dynamical systems. *Europhysics Letters* 4: 324-327.

Faure F, Korn H (1998). A new method to estimate the Kolmogorov entropy from recurrence plots: its application to neuronal signals. *Physica D* 122: 265-279.

Filligoi G, Felici F (1999). Detection of hidden rhythms in surface EMG signals with a non-linear time-series tool. *Medical Engineering & Physics* 21: 439-448.

Frontali C, Pizzi E (1999). Similarity in oligonucleotide usage in introns and intergenic regions contributes to long-range correlation in the Caenorhabditis elegans genome. *Gene* 232: 87-95.

Gardner M (1978). White and brown music, fractal curves and one-over-f noise. *Scientific American* 238: 16-32 (April).

Gao J, Cai H (2000). On the structures and quantification of recurrence plots. *Physics Letters A*, 270: 75-87.

Giuliani A, Benigni R, Sirabella P, Zbilut JP Colosimo A (2000). Exploiting the information content of protein sequences using time-series methods: a case study in rubredoxins. *Biophysical Journal* 78: 136-149.

Giuliani A, Benigni B, Zbilut J, Webber CL, Sirabella P, Colosimo P (2002). Nonlinear signal analysis methods in the elucidation of protein sequence/structure relationships. *Chemical Reviews* 102(5): 1471-1492.

Giuliani A, Colafranceschi M, Webber Jr CL, Zbilut JP (2001). A complexity score derived from principal components analysis of nonlinear order measures. *Physica A* 301: 567-588.

Giuliani A, Colosimo A, Benigni R, Zbilut JP (1998). On the constructive role of noise in spatial systems. *Physics Letters A* 247: 47-52.

Haga SB, Venter JC (2003). FDA races in wrong direction. *Science* 301: 466.

Huberman B (1989). The collective brain. *Int J of Neural Systems* 1: 41-45.

Ikagawa S, Shinohara M, Fukunaga T, Zbilut JP, Webber, Jr CL (2000). Nonlinear time-course of lumbar muscle fatigue using recurrence quantifications. Biological Cybernetics 82: 373-382.

Laughlin RB, Pines D, Schmalian J, Stojkovic BP, Wolynes P (2000). The middle way. *Proceedings of the National Academy of Sciences* 97: 32-37.

Manetti C, Ceruso M-A, Giuliani A, Webber, Jr CL, Zbilut JP (1999). Recurrence quantification analysis as a tool for the characterization of molecular dynamics simulations. *Physical Review E* 59: 992-998.

Manetti C, Ceruso M-A, Giuliani A, Webber CL, Zbilut JP (1999). Recurrence quantification analysis in molecular dynamics. *Annals of the New York Academy of Sciences* 879: 258-266.

Manetti C, Giuliani A, Ceruso M-A, Webber, Jr CL, Zbilut JP (2001). Recurrence Analysis of Hydration Effects on Nonlinear Protein Dynamics: Multiplicative Scaling and Additive Processes. *Physics Letters A* 281: 317-323.

Marino AA, Nilsen E, Frilot C (2002). Consistent magnetic-field induced dynamical changes in rabbit brain activity detected by recurrence quantification analysis. *Brain Research* 951: 301-310.

Marwan N, Kurths J (2002). Nonlinear analysis of bivariate data with cross recurrence plots. *Physics Letters A* 302: 299-307.

Masia M, Bastianoni S, Rustici M (2001). Recurrence quantification analysis of spatio-temporal chaotic transient in a closed unstirred Belousov-Zhabotinsky reaction. *Physical Chemistry Chemical Physics* 3: 245516-245520.

McCulloch WS, Pitts WS (1943). A logical calculus of ideas immanent in nervous activity. *Bull Math Biophys* 5: 115-133.

McGuire G, Azar NB, Shelhamer M (1997). Recurrence matrices and the preservation of dynamical properties, *Physics Letters A* 237: 43-47.

Montroll EW, Shlesinger MF (1982). On 1/f noise and other distributions with long tails. *Proceedingss of the National Academy of Sciences* 79: 3380-3383.

Orsucci F, Walter K, Giuliani A, Webber Jr CL, Zbilut JP (1999). Orthographic structuring of human speech and texts: linguistic application of recurrence quantification analysis. *Int J Chaos Theory Applications* 4:29-38.

Penrose R (1974). The role of aesthetics in pure and applied mathematical research. *Bull Inst Math & Its Appl* 10: 266-271.

Rapp PE (1993). Chaos in the neurosciences: Cautionary tales from the frontier. *The Biologist* (Institute of Biology, London) 40: 89-94.

Riley MA, Balasubramaniam R, Turvey MT (1999). Recurrence quantification analysis of postural fluctuations. *Gait & Posture* 9:

65-78.

Ruelle D (1994). Where can one hope to profitably apply the ideas of chaos? *Physics Today* 47 (July): 24-30.

Shockley K, Butwill M, Zbilut JP, Webber Jr CL (2002). Cross recurrence quantification of coupled oscillators. *Physics Letters A* 305:59-69

Sirabella P, Giuliani A, Zbilut J, Colosimo A (2001). Recurrence quantification analysis and multivariate statistical methods in the study of protein sequences. *Recent Res Devel Protein Eng* 1: 261-275.

Stelzel W, Kautzky T, Gaitzsch P, Hubler A, Lüscher E (1988). Über die Eindeutigkeit der Lösungen der Eulerschen Gleichungen in der klassischen Mechanik, *Helv Phys Acta* 61: 224-227.

Strozzi F, Zaldivar-Comenges J-M, Zbilut JP (2002). Application of nonlinear time series analysis techniques to high frequency currency exchange data. *Physica A* 312: 520-538.

Thiel M, Romano M, Kurths J, Meucci R, Allaria E, Arecchi FT (2002). Influence of observational noise on the recurrence quantification analysis, *Physica D* 171: 138-152.

Thorburn WM (1918). The myth of Occam's razor. *Mind* 217: 345-353.

Thomasson N, Hoeppner TJ, Webber, Jr CL, Zbilut JP (2001). Recurrence quantification in epileptic EEGs. *Physics Letters A* 279: 94-101.

Thomasson N, Webber, CL Jr, Zbilut JP (2002). Application of recurrence quantification analysis to EEG signals. *Int Journal for Computers and their Applications* 9: 1-6.

Trulla LL, Giuliani A, Zbilut JP, Webber Jr CL (1996). Recurrence quantification analysis of the logistic equation with transients. *Physics Letters A* 223: 225-260.

Webber Jr CL, Zbilut JP (1994). Dynamical assessment of physiological systems and states using recurrence plot strategies. *Journal of Applied Physiology* 76: 965-973.

Webber Jr CL, Giuliani A, Zbilut JP, Colosimo A (2001). Elucidating protein secondary structures using alpha-carbon recurrence quantifications. *Proteins Structure, Function, and Genetics* 44: 292-303.

Zbilut JP, Webber Jr C (1992). Embeddings and delays as derived from quantification of recurrence plots. *Physics Letters A* 171: 199-203.

Zbilut JP, Dixon DD, Zak M (2002). Detecting singularities of piecewise deterministic (terminal) dynamics in experimental data. *Physics Letters A* 304: 95-101.

Zbilut JP, Giuliani A, Webber Jr CL (1998). Recurrence quantification analysis and principal components in the detection of short complex signals. *Physics Letters A* 237: 131-135.

Zbilut JP, Giuliani A, Webber, Jr CL (1998). Detecting deterministic signals in exceptionally noisy environments using cross recurrence quantification. *Physics Letters A* 246: 122-128.

Zbilut JP, Giuliani A, Webber, Jr CL (2000). Recurrence quantification analysis as an empirical test to distinguish relatively short deterministic versus random number series. *Physics Letters A* 267: 174-178.

Zbilut JP, Giuliani A, Webber, Jr. CL, Colosimo A (1998). Recurrence quantification analysis in structure function relationships of proteins: An overview of a general methodology applied to the case of TEM-1 Beta -Lactamase. *Protein Engineering* 11 (2): 87-93.

Zbilut JP, Sirabella P, Giuliani A, Manetti C, Colosimo A, Webber, Jr CL (2002). Review of nonlinear analysis of proteins through recurrence quantification. *Cell Biochemistry and Biophysics* 36: 67-87.

Zbilut JP, Thomasson N, Webber Jr CL (2002). Recurrence quantification analysis as a tool for nonlinear exploration of nonstationary cardiac signals. *Medical Engineering and Physics* 24: 53-60.

Zbilut JP, Webber Jr CL, Colosimo A (2000). The role of hydrophobicity patterns in prion folding as revealed by recurrence quantification analysis of primary structure. *Protein Engineering* 13: 99-104.

Zbilut JP, Zak M, Meyers Ronald E (1996). A terminal dynamics model of the heartbeat. *Biological Cybernetics* 75: 277-280.

Zbilut JP, Zak M, Webber Jr CL (1995). Physiological singularities in respiratory and cardiac dynamics. *Chaos, Solitons and Fractals* 5: 1509-1516.

Zbilut JP, Zaldivar-Comenges J-M, Strozzi F (2002). Recurrence quantification based-Liapunov exponents for monitoring divergence in experimental data. *Physics Letters A* 297: 173-181.

Zimatore G, Giuliani A, Parlapiano C, Grisanti G, Colosimo A (2000). Revealing deterministic structures in click-evoked otoacoustic emissions. *Journal of Applied Physiology* 88: 1431-1437.

Index

A

abstraction · 105, 139
adaptability · 6, 48, 76, 86, 89, 106, 111, 135
analysis · viii, 9, 11, 35, 42, 47, 59, 63, 68, 73, 74, 81, 83, 84, 87, 91, 92, 93, 95, 109, 110, 115, 117, 119, 120, 126, 139, 144, 145, 148, 149, 160, 171, 173, 176, 177, 178, 187, 198, 204, 223, 224, 225, 226, 227, 228, 229
arationality · 100
Aristotle · 7, 75, 166, 170, 173
arm · 22, 24, 41, 42, 109, 110, 111, 112, 198, 200, 201, 213
art · 18, 19, 143, 150, 151, 152, 153, 155, 156, 157, 158, 178
attractor · 8, 12, 13, 14, 16, 19, 20, 23, 24, 26, 27, 28, 29, 30, 32, 59, 60, 62, 63, 88, 89, 90, 92, 98, 100, 102, 103, 104, 106, 107, 108, 110, 122, 124, 126, 139, 149, 159, 161, 169, 179, 181, 182, 199, 205

B

Bak · 161, 221, 224
basin · 8, 13, 31, 52
basketball · 107, 161
bias · 95, 140
biology · ix, 4, 5, 16, 35, 49, 73, 74, 126, 134, 137, 144, 178
Boltzmann · 3
brain · 6, 17, 24, 32, 73, 80, 89, 90, 91, 92, 93, 98, 99, 100, 101, 103, 104, 105, 106, 107, 108, 109, 139, 154, 155, 159, 160, 170, 171, 172, 173, 175, 226, 227
Buridan · 7

C

Cervantes · 153

chaos · vii, 6, 8, 15, 18, 26, 27, 28,
  35, 45, 46, 47, 48, 49, 53, 59, 60,
  61, 62, 63, 64, 67, 68, 74, 75, 77,
  78, 85, 89, 91, 99, 100, 101, 102,
  103, 109, 122, 135, 136, 143,
  144, 165, 166, 169, 170, 179,
  181, 194, 224, 228
chaotic · 4, 6, 9, 15, 18, 20, 22,
  27, 30, 31, 35, 37, 38, 46, 48, 49,
  59, 60, 61, 64, 66, 73, 74, 75, 78,
  84, 86, 89, 102, 104, 105, 110,
  111, 132, 135, 140, 144, 165,
  166, 169, 172, 187, 189, 194,
  198, 215, 225, 227
chemical · ix, 47, 99, 112, 114,
  117, 204
circulation · 79
cluster · 145
Coddington · 221
cognition · 102, 108, 143, 152,
  155, 158
cognitive · 100, 104, 105, 143,
  149, 151, 152, 157, 158, 173
coherence · 102, 155
collective · 5, 17, 100, 101, 102,
  106, 107, 108, 109, 226
compartment · viii, 126, 130, 132,
  134
condition · 17, 18, 20, 25, 26, 27,
  28, 29, 39, 53, 74, 76, 89, 99,
  101, 103, 108, 121, 131, 156,
  182, 184, 205, 210
control · 4, 5, 32, 36, 42, 48, 49,
  58, 64, 66, 67, 68, 69, 78, 79, 80,
  85, 88, 93, 95, 96, 97, 100, 106,
  107, 108, 109, 111, 117, 130,
  144, 162, 166, 179, 208, 209,
  210, 211
convergence · 15, 29, 30
coordinate · 14, 43, 50, 52, 175
creativity · 6, 9, 18, 98, 100,
  101, 102, 106, 137, 156
Crutchfield · 16, 19, 68, 225
Csikszentmihalyi · 150, 155,
  156, 222

D

degrees of freedom · 36, 58,
  102, 135, 143, 175, 195
derivative · vii, viii, 11, 25, 61,
  111, 126, 183, 204, 208, 210,
  211
detection · 35, 94, 111, 204, 229
determinism · 2, 6, 11, 17, 28, 32,
  33, 38, 39, 45, 46, 47, 53, 55, 62,
  67, 68, 75, 76, 77, 79, 92, 116,
  117, 118, 119, 122, 133, 135,
  136, 141, 144, 147, 156, 165,
  166, 167, 171, 172, 173, 175,
  178, 190
differentiability · 8, 13
diffusion · 47, 63, 123
dimension · 37, 51, 58, 74, 75, 92,
  118, 136, 175, 184, 195, 208
dissipation · 16, 25, 26, 29
divergence · 15, 16, 24, 25, 27,
  30, 39, 62, 75, 83, 110, 111, 112,
  117, 118, 119, 183, 190, 198,
  200, 201, 202, 204, 206, 207,
  208, 210, 230
Dixon · viii, ix, 50, 131, 208, 225,
  229
Dobson · 225
dynamics · vii, viii, ix, 4, 6, 7, 8, 9,
  11, 13, 15, 16, 17, 19, 20, 21, 22,
  23, 24, 25, 26, 27, 28, 29, 30, 31,
  32, 33, 35, 36, 37, 38, 39, 40, 42,
  44, 47, 48, 49, 51, 53, 57, 59, 60,
  62, 63, 67, 73, 74, 75, 76, 77, 78,
  79, 81, 83, 84, 86, 87, 88, 89, 92,
  94, 98, 99, 101, 102, 103, 104,
  105, 106, 107, 108, 109, 110,
  111, 112, 116, 121, 122, 132,
  133, 134, 135, 136, 137, 141,
  144, 145, 147, 149, 158, 159,
  160, 161, 165, 166, 167, 169,
  170, 171, 172, 173, 174, 181,
  182, 183, 184, 195, 196, 197,
  199, 200, 207, 208, 210, 213,
  214, 215, 223, 224, 225, 226,
  227, 229, 230

E

Eckmann · 190, 225
economics · 143, 144
Economics · 143, 144
Einstein · 2, 5, 7, 49, 170, 222
electrocardiogram · 171
electroencephalogram · 103, 171
energy · 20, 26, 28, 30, 40, 49, 82, 88, 99, 116, 122, 144, 152, 171, 172, 176, 204, 205
entropy · 2, 31, 74, 106, 135, 176, 187, 190, 193, 225
epilepsy · 90, 91, 93, 98
equilibrium · 13, 18, 21, 24, 25, 26, 27, 28, 29, 31, 37, 50, 62, 68, 89, 101, 103, 108, 182, 205
exchange rate · 149
Exchange rate · 149

F

feedback · 73, 79, 91, 133
fibrillation · 189
filament · 20, 30
filtering · 75, 83, 111, 199, 200, 202
first passage time · 43
Fisher · 3
flexibility · 3, 66, 106, 121, 165
fluctuation · 194, 215
force · 22, 25, 26, 31, 36, 42, 115, 147, 153, 158
friction · 13, 25, 26, 29

G

generalization · 100, 102, 103, 104, 105, 108, 122

Giuliani · ix, 224, 226, 227, 228, 229, 230

H

Hamiltonian · 171
historicism · 156, 157, 158, 159, 160

I

inertia · 6, 49, 157
informatics · 114
information · 2, 6, 15, 16, 17, 19, 28, 43, 60, 61, 62, 63, 65, 89, 92, 93, 98, 99, 100, 101, 102, 103, 106, 108, 109, 111, 117, 146, 162, 173, 175, 176, 178, 184, 187, 190, 196, 213, 226
instability · vii, 5, 7, 8, 15, 16, 18, 19, 22, 24, 30, 31, 49, 60, 62, 68, 101, 143, 210, 211, 223, 225
insulin · 131, 132
intelligence · 100, 102, 106, 107, 108, 109, 172, 223, 224
invariant · 17, 35, 171, 172
irrationality · 6, 100
irreversibility · 26, 31, 98, 101

K

kinetics · 5, 88, 114, 127

L

Lagrangian · 28, 172
Laplace · 127, 128, 130, 172, 178
Liapunov exponent · 15, 16, 24, 27, 35, 37, 38, 60, 61, 63, 74, 75,

83, 84, 92, 110, 112, 144, 146,
172, 187, 190, 193, 198, 210,
214, 215, 230
limit cycle · 37, 42, 80, 169
Lipschitz · 13, 17, 18, 20, 21, 22,
24, 25, 26, 27, 28, 29, 39, 53, 76,
77, 81, 84, 87, 89, 92, 99, 101,
102, 103, 108, 121, 132, 134,
135, 166, 174, 180, 181, 182
Lipschitz condition · 13, 17,
18, 20, 25, 26, 27, 28, 29, 39, 76,
77, 81, 89, 99, 101, 102, 103,
108, 132, 166, 180, 182
Lovejoy · 140, 222

M

manager · 162
mechanics · ix, 2, 4, 5, 11, 28, 31,
41, 46, 68, 103, 109, 170, 172,
181, 221
modeling · 19, 25, 28, 36, 46, 98,
99, 101, 103, 108, 113, 179, 183,
213
motion · 4, 12, 16, 17, 19, 22, 25,
26, 27, 28, 30, 33, 39, 41, 42, 44,
45, 47, 50, 52, 55, 60, 62, 64, 68,
77, 78, 89, 101, 102, 110, 111,
134, 151, 152, 198, 200, 201,
213, 224, 225
Mozart · 156
music · 70, 152, 156, 226

N

neural net · 16, 17, 90, 98, 99,
100, 101, 102, 103, 104, 161
neural network · 16, 98, 99,
100, 101, 102, 103, 104, 161
neurodynamics · 32, 89, 98, 101,
102, 103, 104, 105, 106, 108,
109
neuron · 17, 102, 104, 173

Newton · 2, 11, 25, 28, 37, 38, 44,
45, 46, 170, 171, 178
noise · viii, 2, 3, 8, 20, 22, 23, 35,
36, 37, 38, 40, 42, 43, 46, 47, 48,
49, 53, 55, 57, 58, 62, 63, 64, 67,
68, 70, 75, 77, 80, 82, 83, 88, 89,
92, 93, 94, 95, 96, 97, 99, 111,
132, 133, 134, 135, 136, 143,
144, 145, 151, 155, 158, 165,
167, 179, 183, 187, 195, 196,
198, 204, 208, 210, 211, 214,
215, 224, 225, 226, 227, 228
nondeterminism · viii, 11, 18, 21,
22, 33, 38, 39, 40, 42, 43, 44, 47,
52, 53, 55, 57, 60, 62, 63, 64, 66,
67, 68, 76, 77, 78, 79, 80, 81, 83,
90, 99, 103, 106, 108, 122, 132,
135, 141, 162, 166, 171, 174,
181, 197, 200, 224
nondeterministic · viii, 11, 18,
21, 22, 33, 38, 39, 40, 42, 43, 44,
47, 52, 53, 55, 57, 60, 62, 63, 64,
66, 67, 68, 76, 77, 78, 79, 80, 81,
83, 90, 99, 103, 106, 108, 122,
132, 135, 141, 162, 166, 171,
174, 181, 197, 200, 224
nonequilibrium · 18, 101
Nyquist · 70, 183, 197

O

oscillation · 185, 212

P

parallel · 7, 17, 21, 43, 49, 79,
80, 100, 102, 184, 194, 204
pattern · 98, 106, 111, 118, 121,
170, 172, 175, 213
pause · 42, 86, 87, 110, 111, 200,
213
pdf · 159, 161, 176
Pearson · 3

Index 235

perception · 1, 47, 67, 104, 140, 143, 151, 152, 154, 155, 158, 172
periodic · 14, 15, 16, 19, 41, 42, 64, 74, 78, 89, 100, 104, 112, 122, 123, 166, 183, 187, 194, 197, 225
personality · 158, 159, 160, 172
phase space · 21, 24, 43, 47, 50, 51, 52, 53, 55, 58, 59, 60, 62, 63, 66, 75, 77, 78, 81, 104, 110, 132, 135, 175, 196, 199, 205, 206
physical model · 78, 102, 108
Picasso · 151, 154, 155, 157
piecewise determinism · 136
point · 2, 5, 6, 7, 8, 11, 22, 23, 24, 26, 27, 28, 29, 30, 32, 38, 40, 43, 48, 50, 51, 52, 53, 55, 58, 60, 61, 63, 66, 67, 68, 70, 75, 76, 77, 78, 82, 93, 104, 110, 111, 114, 115, 116, 117, 130, 131, 132, 135, 144, 148, 151, 158, 159, 165, 166, 172, 177, 178, 182, 183, 184, 191, 200, 204, 205, 207, 208, 209, 211, 213, 214
power spectrum · 91, 176, 187
predictability · 6, 20, 24, 30, 62, 63, 213, 223
pressure · 80, 81, 224
Prigogine · 6, 13, 17, 31, 222
probability · vii, 2, 3, 22, 23, 24, 27, 28, 31, 32, 45, 55, 62, 63, 64, 78, 88, 94, 105, 107, 108, 109, 122, 123, 124, 137, 139, 141, 145, 148, 156, 159, 160, 162, 167, 178, 181
psychology · 49, 67, 143, 158, 159

R

random · vii, 3, 5, 6, 18, 19, 22, 24, 27, 30, 31, 46, 48, 53, 55, 62, 67, 74, 75, 86, 101, 102, 103, 105, 107, 108, 121, 132, 136, 137, 138, 139, 144, 156, 161, 165, 170, 173, 174, 176, 177, 178, 184, 185, 190, 194, 229
randomness · vii, 5, 18, 22, 24, 31, 46, 102, 136, 165, 171, 174, 184
reaction · 113, 204, 205, 209, 210, 211, 212, 216, 227
recurrence · 9, 15, 81, 86, 87, 88, 92, 93, 94, 95, 96, 97, 110, 116, 119, 121, 122, 145, 149, 176, 177, 178, 183, 184, 185, 187, 188, 189, 190, 191, 192, 193, 196, 198, 208, 212, 215, 217, 223, 225, 226, 227, 228, 229
recurrence analysis · 15, 95, 178, 223
recurrence quantification · 9, 81, 92, 110, 116, 198, 226, 227, 228, 229
regular · 22, 26, 27, 29, 37, 73, 93, 111, 182
relaxation · 28, 98, 182
repeller · 12, 26, 27, 29, 30, 182
Ruelle · 15, 170, 225, 228

S

Salieri · 156
sampling · 70, 74, 82, 94, 95, 110, 111, 183, 190, 197, 198, 199, 200
Sartre · 159, 222
scaling · 54, 79, 92, 93, 94, 95, 96, 98, 119, 145, 173, 177, 187, 196, 197
self · 24, 32, 33, 108, 113, 151, 156, 159, 160, 161, 177, 179
self-organized · 161, 179
sensitivity · 18, 27, 204, 209
singularity · 6, 7, 8, 20, 21, 23, 24, 25, 29, 41, 44, 47, 52, 53, 55, 57, 60, 61, 62, 63, 65, 66, 68, 77, 78, 79, 80, 81, 87, 88, 110, 111, 112, 116, 117, 118, 119, 121, 122, 124, 132, 135, 149, 152,

158, 160, 161, 165, 166, 178,
181, 182, 183, 198, 199, 212,
213, 217
sociology · 143
Spanish steps · ix, 151, 152, 153
spectrum · 91, 173, 175, 187
stability · 8, 11, 13, 14, 15, 18,
48, 89, 98, 111, 169, 172, 179,
215
state · 3, 7, 12, 19, 25, 36, 38, 46,
47, 53, 62, 76, 98, 104, 109, 116,
119, 132, 152, 170, 171, 172,
173, 175, 179, 195, 207, 215
static · 13, 14, 15, 19, 26, 29,
104, 113
stationarity · 15, 36, 47, 75, 92,
126, 149, 178, 196
statistics · 2, 3, 60, 62, 68, 137,
190
stochastic · vii, ix, 2, 3, 22, 23,
24, 31, 32, 37, 55, 64, 74, 75, 85,
86, 88, 89, 90, 92, 94, 103, 104,
105, 106, 107, 108, 122, 123,
124, 136, 139, 147, 148, 155,
156, 158, 159, 160, 161, 172,
181
stochastic attractor · 23, 24,
32, 88, 89, 90, 92, 103, 104, 105,
106, 107, 108, 122, 124, 126,
139, 149, 155, 156, 159, 160,
161, 181
stochastic process · 31, 88,
103, 106, 107, 108, 122, 123,
147, 159, 160, 181
stochasticity · 7, 8, 67, 122,
123, 223
stock market · 145
superposition · 81, 114, 173
system · 4, 5, 8, 12, 13, 14, 15, 17,
18, 19, 20, 25, 26, 27, 28, 31, 32,
33, 35, 37, 38, 39, 41, 42, 43, 44,
45, 46, 48, 49, 50, 52, 53, 57, 58,
59, 60, 61, 62, 64, 66, 67, 68, 74,
75, 76, 77, 78, 79, 80, 81, 83, 86,
88, 89, 91, 92, 95, 98, 99, 100,
101, 102, 103, 104, 106, 107,

108, 115, 119, 122, 127, 130,
131, 132, 133, 134, 144, 149,
151, 160, 166, 169, 170, 171,
172, 173, 175, 176, 177, 178,
179, 181, 182, 183, 184, 190,
194, 195, 204, 205, 207, 208,
215, 225

T

Taylor series · 25, 58, 59
terminal · 21, 22, 23, 26, 27, 28,
29, 30, 31, 32, 86, 98, 99, 100,
101, 102, 103, 105, 106, 107,
108, 109, 181, 229, 230
Thorburn · 228
time series · 11, 14, 29, 48, 53,
61, 74, 75, 80, 83, 84, 91, 93, 95,
96, 109, 110, 111, 112, 115, 144,
146, 149, 165, 171, 176, 178,
184, 189, 190, 195, 198, 200,
202, 207, 210, 212, 215, 228
transients · 26, 28, 37, 52, 182,
228
Trulla · 228

U

uniqueness · 13, 17, 18, 20, 21,
24, 26, 27, 28, 31, 39, 47, 50, 76,
77, 99, 166, 181, 205
universal · 19, 45, 47, 77, 103,
105, 108
unpredictability · 8, 18, 27,
30, 53, 62, 101

V

Venter · 137, 226
volume · 16, 58, 59, 60, 63, 90,
110, 121, 175, 199, 206, 207,
214

W

wave · 20, 30, 44, 86, 87, 88, 109, 110, 122, 186, 188, 198, 199, 200
wavelet · 176
Webber · ix, 190, 223, 224, 226, 227, 228, 229, 230
William of Occam · 2, 7

Winfree · 6, 222

Z

Zak · vii, ix, 223, 224, 229, 230
Zbilut · vii, 190, 223, 224, 225, 226, 227, 228, 229, 230
Zimatore · 230